T5-CVD-939

RUU660 26690

Worked examples in the
GEOMETRY OF
CRYSTALS

H K D H BHADESHIA

Lecturer
in the Department of
Materials Science and Metallurgy
University of Cambridge

The Institute of Metals

1987

REF
QD
911
.B48
1987

Book 377
ISBN 0 904357 94 5

published in 1987 by

The Institute of Metals
1 Carlton House Terrace   and
London SW1Y 5DB

The Institute of Metals
North American Publications Center
Old Post Road
Brookfield, VT 05036, USA

© THE INSTITUTE OF METALS 1987

ALL RIGHTS RESERVED

British Library Cataloguing in Publication Data

Bhadeshia, H.K.D.H.
    Worked examples in the geometry of crystals.
    1. Crystallography, Mathematical ——
    Problems, exercises, etc.
    I. Title
    548'.1   QD911

    ISBN 0-904357-94-5

Library of Congress Cataloging in Publication Data

applied for

COVER ILLUSTRATION
shows a net-like sub-grain boundary
in annealed bainite; x 150 000
(Photograph by courtesy of J R Yang)

Compiled from original typesetting and
illustrations provided by the author

Printed and made in England by
The Chameleon Press Ltd, London SW18 4SG

WORKED EXAMPLES IN THE GEOMETRY OF CRYSTALS

H. K. D. H. Bhadeshia

PREFACE

A large part of crystallography deals with the way in which atoms are arranged in single crystals. On the other hand, a knowledge of the relationships between crystals in a polycrystalline material can be fascinating from a materials science point of view, and it is this aspect of crystallography which is the subject of this monograph. The monograph is aimed at both undergraduates and graduate students and assumes only an elementary knowledge of crystallography. Although use is made of vector and matrix algebra, readers not familiar with these methods should not be at a disadvantage after studying appendix 1. In fact, the mathematics necessary for a good grasp of the subject is not very advanced but the concepts involved can be difficult to absorb. It is for this reason that the book is based on worked examples, which are intended to make the ideas less abstract.

Due to its wide-ranging applications, the subject has developed with many different schemes for notation and this can be confusing to the novice. The extended notation used throughout this text was first introduced by Mackenzie and Bowles; I believe that this is a clear and unambiguous scheme which is particularly powerful in distinguishing between representations of deformations and axis transformations.

The monograph begins with an introduction to the range of topics that can be handled using the concepts developed in detail in later chapters. The introduction also serves to familiarise the reader with the notation used. The other chapters cover orientation relationships, aspects of deformation, martensitic transformations and interfaces.

In preparing this book, I have benefited from the support of Professor R. W. K. Honeycombe, Professor D. Hull, Dr. F. B. Pickering and Dr. J. Wood. I am especially grateful to Professor J. W. Christian and Dr. J. F. Knott for their detailed comments on the text, and to many students who have over the years helped clarify my understanding of the subject. It is a pleasure to acknowledge the unfailing support of my family.

August 1986

iii

CONTENTS

INTRODUCTION

Crystallographic analysis, as applied in Metallurgy, can be classified into two main subjects; the first of these has been established ever since it was realised that metals have a crystalline character, and is concerned with the clear description and classification of atomic arrangements. X-ray and electron diffraction methods combined with other structure sensitive physical techniques have successfully been utilised to study the crystalline state, and the information obtained has long formed the basis of metallurgical investigations on the role of the discrete lattice in influencing the behaviour of commonly used engineering materials.

The second aspect, which is the subject of this monograph, is more recent and took off in earnest when it was noticed that accurate experimental data on martensitic transformations showed many apparent inconsistencies. Matrix methods were used in resolving these difficulties, and led to the formulation of the phenomenological theory of martensite[1,2]. Similar methods have since widely been applied in metallurgy; the nature of shape changes accompanying displacive transformations and the interpretation of interface structure are two examples. Despite the apparent diversity of applications, there is a common theme in the various theories, and it is this which makes it possible to cover a variety of topics in this monograph.

Throughout this monograph, every attempt has been made to keep the mathematical content to a minimum and in as simple a form as the subject allows; the student need only have an elementary appreciation of matrices and of vector algebra. Appendix 1 provides a very brief revision of these aspects, together with references to some standard texts available for further consultation.

The purpose of this introductory chapter is to indicate the range of topics that can be tackled using the crystallographic methods, while at the same time familiarising the reader with vital notation; many of the concepts introduced are covered in more detail in the chapters that follow. It is planned to introduce the subject with reference to the martensite transformation in steels, which not only provides a good example of the application of crystallographic methods, but which is also a transformation of major practical importance.

At temperatures between 1185K and 1655K, pure iron exists as a face-centered cubic (FCC) arrangement of iron atoms. Unlike other FCC metals, lowering the temperature leads to the formation of a body-centered cubic (BCC) allotrope of iron. This change in crystal structure can occur in at least two different ways. Given sufficient atomic mobility, the FCC lattice can undergo complete reconstruction into the BCC form, with considerable unco-ordinated diffusive mixing-up of atoms at the transformation interface. On the other hand, if the FCC phase is rapidly cooled to a very low temperature, well below 1185K, there may not be enough time or atomic mobility to facilitate diffusional transformation. The driving force for transformation nevertheless increases with undercooling below 1185K, and the diffusionless formation of BCC martensite eventually occurs, by a displacive or "shear" mechanism, involving the systematic and co-ordinated transfer of atoms across the interface. The formation of this BCC martensite is indicated by a very special change in the shape of the austenite (γ) crystal, a change of shape which is beyond that expected just on the basis of a volume change effect. The nature of this shape change will be discussed later in the text, but for the present it is taken to imply that the transformation from austenite to ferrite occurs by some kind of a deformation of the austenite lattice. It was E. C. Bain [3] who in 1924 introduced the concept that the structural change from austenite to martensite might occur by a homogeneous deformation of the austenite lattice, by some kind of an upsetting process, the so called Bain Strain.

Definition of a Basis

Before attempting to deduce the Bain Strain, we must establish a method of describing the austenite lattice. Fig. 1a shows the FCC unit cell of austenite, with a vector $\underline{u}$ drawn along the cube diagonal. To specify the direction and magnitude of this vector, and to relate it to other vectors, it is necessary to have a reference set of co-ordinates. A convenient reference frame would be formed by the three (right-handed) orthogonal vectors $\underline{a}_1$, $\underline{a}_2$ and $\underline{a}_3$, which lie along the unit cell edges, each of magnitude a_γ, the lattice parameter of the austenite. The term orthogonal implies a set of mutually perpendicular vectors, each of which can be of arbitrary magnitude; if these vectors are mutually perpendicular and of unit magnitude, they are called orthonormal.

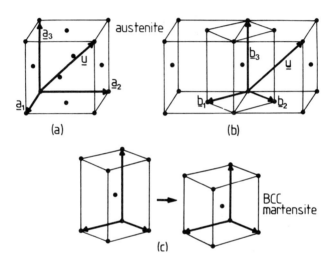

Fig. 1   a) Conventional FCC unit cell b) Relation between FCC and BCT cells of austenite   c) Bain Strain deforming the austenite lattice into a BCC martensite lattice.

The set of vectors $\underline{a}_i$ (i = 1 to 3) are called the basis vectors, and the basis itself may be identified by a basis symbol, 'A' in this instance.

The vector $\underline{u}$ can then be written as a linear combination of the basis vectors:

$$\underline{u} = u_1\underline{a}_1 + u_2\underline{a}_2 + u_3\underline{a}_3,$$

where u_1, u_2 and u_3 are its components, when $\underline{u}$ is referred to the basis A. These components can conveniently be written as a single-row matrix $(u_1\ u_2\ u_3)$ or as a single-column matrix:

$\begin{pmatrix} u_1 \\ u_2 \\ u_3 \end{pmatrix}$ This column representation can, for convenience, be written using square brackets as: $[u_1\ u_2\ u_3]$

It follows from this that the matrix representation of the vector $\underline{u}$ (Fig. 1a), with respect to the basis A is

$$(\underline{u};A) = (u_1\ u_2\ u_3) = (1\ 1\ 1)$$

where $\underline{u}$ is represented as a row vector. $\underline{u}$ can alternatively be represented as a column vector

2

$$[A;\underline{u}] = [u_1\ u_2\ u_3] = [1\ 1\ 1]$$

The row matrix $(\underline{u};A)$ is the transpose of the column matrix $[A;\underline{u}]$, and vice versa. The positioning of the basis symbol in each representation is important as will be seen later. The notation, which is due to Mackenzie and Bowles [2], is particularly good in avoiding confusion between bases.

Co-ordinate Transformations

From Fig. 1a, it is evident that the choice of basis vectors $\underline{a}_i$ is arbitrary though convenient; Fig. 1b illustrates an alternative basis, a body-centered tetragonal (BCT) unit cell describing the same austenite lattice. We label this as basis 'B', consisting of basis vectors $\underline{b}_1$, $\underline{b}_2$ and $\underline{b}_3$ which define the BCT unit cell. It is obvious that $[B;\underline{u}] = [0\ 2\ 1]$, compared with $[A;\underline{u}] = [1\ 1\ 1]$. The following vector equations illustrate the relationships between the basis vectors of A and those of B (Fig. 1):

$$\underline{a}_1 = 1\underline{b}_1 + 1\underline{b}_2 + 0\underline{b}_3$$
$$\underline{a}_2 = \bar{1}\underline{b}_1 + 1\underline{b}_2 + 0\underline{b}_3$$
$$\underline{a}_3 = 0\underline{b}_1 + 0\underline{b}_2 + 1\underline{b}_3$$

These equations can also be presented in matrix form as follows:

$$(\underline{a}_1\ \underline{a}_2\ \underline{a}_3) = (\underline{b}_1\ \underline{b}_2\ \underline{b}_3) \begin{pmatrix} 1 & \bar{1} & 0 \\ 1 & 1 & 0 \\ 0 & 0 & 1 \end{pmatrix}$$

.....(1)

This 3 x 3 matrix representing the co-ordinate transformation is denoted (B J A) and transforms the components of vectors referred to the A basis to those referred to the B basis. The first column of (B J A) represents the components of the basis vector $\underline{a}_1$, with respect to the basis B, and so on.

The components of a vector $\underline{u}$ can now be transformed between bases using the matrix (B J A) as follows:

$$[B;\underline{u}] = (B\ J\ A)[A;\underline{u}] \qquad .....(2a)$$

Notice the juxtapositioning of like basis symbols. If (A J' B) is the transpose of (B J A), then eq.2a can be rewritten as

$$(\underline{u};B) = (\underline{u};A)(A\ J'\ B) \qquad .....(2b)$$

Writing (A J B) as the inverse of (B J A), we obtain:

$$[A;\underline{u}] = (A\ J\ B)[B;\underline{u}] \qquad .....(2c)$$

and

$$(\underline{u};A) = (\underline{u};B)(B\ J'\ A) \qquad .....(2d)$$

It has been emphasized that each column of (B J A) represents the components of a basis vector of A with respect to the basis B (i.e., $\underline{a}_1 = J_{11}\underline{b}_1 + J_{21}\underline{b}_2 + J_{31}\underline{b}_3$ etc.). This procedure is also adopted in (for example) Refs.4,5. Some texts use the convention that each row of (B J A) serves this function (i.e., $\underline{a}_1 = J_{11}\underline{b}_1 + J_{12}\underline{b}_2 + J_{13}\underline{b}_3$ etc.). There are others where a mixture of both methods is used - the reader should be aware of this problem.

3

Example 1: Co-ordinate transformations.

Two adjacent grains of austenite are represented by bases 'A' and 'B' respectively. The base vectors $\underline{a}_i$ of A and $\underline{b}_i$ of B respectively define the FCC unit cells of the austenite grains concerned. The lattice parameter of the austenite is a_γ so that $|\underline{a}_i| = |\underline{b}_i| = a_\gamma$. The grains are orientated such that $[0\ 0\ 1]_A \parallel [0\ 0\ 1]_B$, and $[1\ 0\ 0]_B$ makes an angle of 45° with both $[1\ 0\ 0]_A$ and $[0\ 1\ 0]_A$. Prove that if $\underline{u}$ is a vector such that its components in crystal A are given by $[A;\underline{u}] = [\sqrt{2}\ 2\sqrt{2}\ 0]$, then in the basis B, $[B;\underline{u}] = [3\ 1\ 0]$. Show that the magnitude of $\underline{u}$ (i.e., $|\underline{u}|$) does not depend on the choice of the basis.

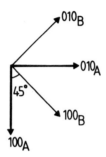

Fig. 2   Diagram illustrating the relation between the bases A and B.

Referring to Fig. 2, and recalling that the matrix (B J A) consists of three columns, each column being the components of one of the basis vectors of A, with respect to B, we have

$[B;\underline{a}_1] = [\cos45 \quad -\sin45 \quad 0]$

$[B;\underline{a}_2] = [\sin45 \quad \cos45 \quad 0]$ and $(B\ J\ A) = \begin{pmatrix} \cos45 & \sin45 & 0 \\ -\sin45 & \cos45 & 0 \\ 0 & 0 & 1 \end{pmatrix}$

$[B;\underline{a}_3] = [\ 0 \qquad\quad 0 \quad\ 1]$

From eq.2a, $[B;\underline{u}] = (B\ J\ A)[A;\underline{u}]$, and on substituting for $[A;\underline{u}] = [\sqrt{2}\ 2\sqrt{2}\ 0]$, we get $[B;\underline{u}] = [3\ 1\ 0]$. Both the bases A and B are orthogonal so that the magnitude of $\underline{u}$ can be obtained using the Pythagoras theorem. Hence, choosing components referred to the basis B, we get:

$$|\underline{u}|^2 = (3|\underline{b}_1|)^2 + (|\underline{b}_2|)^2 = 10a_\gamma^2$$

With respect to basis A,

$$|\underline{u}|^2 = (\sqrt{2}|\underline{a}_1|)^2 + (2\sqrt{2}|\underline{a}_2|)^2 = 10a_\gamma^2$$

Hence, $|\underline{u}|$ is invariant to the co-ordinate transformation. This is a completely general result, since a vector is a physical entity, whose magnitude and direction clearly cannot depend on the choice of a reference frame, a choice which is after all arbitrary.

We note that the components of (B J A) are the cosines of angles between $\underline{b}_i$ and $\underline{a}_j$ and that $(A\ J'\ B) = (A\ J\ B)^{-1}$; a matrix with these properties is described as orthogonal (see appendix). An

4

orthogonal matrix represents an axis transformation between like orthogonal bases.

The Reciprocal Basis

The reciprocal lattice that is so familiar to crystallographers also constitutes a special co-ordinate system, designed originally to simplify the study of diffraction phenomena. If we consider a lattice, represented by a basis symbol A and an arbitrary set of basis vectors $\underline{a}_1$, $\underline{a}_2$ and $\underline{a}_3$, then the corresponding reciprocal basis A^* has basis vectors $\underline{a}_1^*$, $\underline{a}_2^*$ and $\underline{a}_3^*$, defined by the following equations:

$$\underline{a}_1^* = (\underline{a}_2 \wedge \underline{a}_3) / (\underline{a}_1 . \underline{a}_2 \wedge \underline{a}_3) \qquad .....(3a)$$

$$\underline{a}_2^* = (\underline{a}_3 \wedge \underline{a}_1) / (\underline{a}_1 . \underline{a}_2 \wedge \underline{a}_3) \qquad .....(3b)$$

$$\underline{a}_3^* = (\underline{a}_1 \wedge \underline{a}_2) / (\underline{a}_1 . \underline{a}_2 \wedge \underline{a}_3) \qquad .....(3c)$$

In eq.3a, the term $(\underline{a}_1 . \underline{a}_2 \wedge \underline{a}_3)$ represents the volume of the unit cell formed by $\underline{a}_i$, while the magnitude of the vector $(\underline{a}_2 \wedge \underline{a}_3)$ represents the area of the $(1\ 0\ 0)_A$ plane (see appendix). Since $(\underline{a}_2 \wedge \underline{a}_3)$ points along the normal to the $(1\ 0\ 0)_A$ plane, it follows that $\underline{a}_1^*$ also points along the normal to $(1\ 0\ 0)_A$ and that its magnitude $|\underline{a}_1^*|$ is the reciprocal of the spacing of the $(1\ 0\ 0)_A$ planes (Fig. 3).

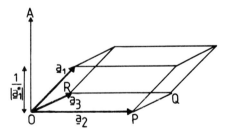

Fig. 3   The relationship between $\underline{a}_1^*$ and $\underline{a}_i$. The vector $\underline{a}_1^*$ lies along the direction OA and the volume of the parallelepiped formed by the basis vectors $\underline{a}_i$ is given by $\underline{a}_1 . \underline{a}_2 \wedge \underline{a}_3$, the area OPQR being equal to $|\underline{a}_2 \wedge \underline{a}_3|$.

The reciprocal lattice is useful in crystallography because it has this property; the components of any vector referred to the reciprocal basis represent the Miller indices of a plane whose normal is along that vector, with the spacing of the plane given by the inverse of the magnitude of that vector. For example, the vector $(\underline{u};A^*) = (1\ 2\ 3)$ is normal to planes with Miller indices $(1\ 2\ 3)$ and interplanar spacing $1/|\underline{u}|$. Throughout this text, the presence of an asterisk indicates reference to the reciprocal basis. Wherever possible, plane normals will be written as row vectors, and directions as column vectors.

We see from eq.3 that

$$\underline{a}_i.\underline{a}_j^* = 1 \text{ when } i = j, \text{ and } \underline{a}_i.\underline{a}_j^* = 0 \text{ when } i \neq j$$

or in other words,

$$\underline{a}_i.\underline{a}_j^* = \delta_{ij} \qquad \qquad .....(4a)$$

where δ_{ij} is the Kronecker delta, which has a value of unity when $i = j$ and is zero when $i \neq j$ (see appendix).

Emphasizing the fact that the reciprocal lattice is really just another convenient co-ordinate system, a vector $\underline{u}$ can be identified by its components $[A;\underline{u}] = [u_1 \ u_2 \ u_3]$ in the direct lattice or $(\underline{u}:A^*) = (u_1^* \ u_2^* \ u_3^*)$ in the reciprocal lattice. The components are defined as usual, by the equations:

$$\underline{u} = u_1\underline{a}_1 + u_2\underline{a}_2 + u_3\underline{a}_3 \qquad \qquad .....(4b)$$

$$\underline{u} = u_1^*\underline{a}_1^* + u_2^*\underline{a}_2^* + u_3^*\underline{a}_3^* \qquad \qquad .....(4c)$$

The magnitude of $\underline{u}$ is given by

$$|\underline{u}|^2 = \underline{u}.\underline{u}$$

$$= (u_1\underline{a}_1 + u_2\underline{a}_2 + u_3\underline{a}_3).(u_1^*\underline{a}_1^* + u_2^*\underline{a}_2^* + u_3^*\underline{a}_3^*)$$

using eq.4a, it is evident that

$$|\underline{u}|^2 = (u_1 u_1^* + u_2 u_2^* + u_3 u_3^*)$$

$$= (\underline{u};A^*)[A;\underline{u}]. \qquad \qquad .....(4c)$$

This is an important result, since it gives a new interpretation to the scalar, or "dot" product between any two vectors $\underline{u}$ and $\underline{v}$ since

$$\underline{u}.\underline{v} = (\underline{u};A^*)[A;\underline{v}] = (\underline{v};A^*)[A;\underline{u}] \qquad \qquad .....(4d)$$

Homogeneous Deformations

We can now return to the question of martensite, and how a homogeneous deformation might transform the austenite lattice (parameter a_γ) to a BCC martensite (parameter a_α). Referring to Fig. 1, the basis 'A' is defined by the basis vectors $\underline{a}_i$, each of magnitude a_γ and basis 'B' is defined by basis vectors $\underline{b}_i$ as illustrated in Fig. 1b. Focussing attention on the BCT representation of the austenite unit cell (Fig. 1b), it is evident that a compression along the $[0 \ 0 \ 1]_B$ axis, coupled with expansions along $[1 \ 0 \ 0]_B$ and $[0 \ 1 \ 0]_B$ would accomplish the transformation of the BCT austenite unit cell into a BCC α cell. This deformation, referred to the basis B, can be written as:

$$\eta_1 = \eta_2 = \sqrt{2}(a_\alpha/a_\gamma)$$

along $[1 \ 0 \ 0]_B$ and $[0 \ 1 \ 0]_B$ respectively and

$$\eta_3 = (a_\alpha/a_\gamma)$$

6

along the $[0\ 0\ 1]_B$ axis.

The deformation just described can be written as a 3x3 matrix, referred to the austenite lattice. In other words, we imagine that a part of a single crystal of austenite undergoes the prescribed deformation, allowing us to describe the strain in terms of the remaining (and undeformed) region, which forms a fixed reference basis. Hence, the deformation matrix does not involve any change of basis, and this point is emphasized by writing it as (A S A), with the same basis symbol on both sides of S:

$$[A;\underline{v}] = (A\ S\ A)[A;\underline{u}] \qquad .....(5)$$

where the homogeneous deformation (A S A) converts the vector $[A;\underline{u}]$ into a new vector $[A;\underline{v}]$, with $\underline{v}$ still referred to the basis A. The difference between a co-ordinate transformation (B J A) and a deformation matrix (A S A) is illustrated in Fig. 4 below, where $\underline{a}_i$ and $\underline{b}_i$ are the basis vectors of the bases A and B respectively.

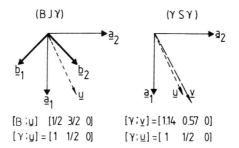

Fig. 4   Difference between co-ordinate transformation and deformation matrix.

We see that a major advantage of the Mackenzie-Bowles notation is that it enables a clear distinction to be made between 3x3 matrices which represent changes of axes and those which represent physical deformations referred to one axis system.

The following additional information can be deduced from Fig. 1:

Vector components before Bain strain	*Vector components after Bain strain*
$[1\ 0\ 0]_A$	$[\eta_1\ 0\ \ \ 0]_A$
$[0\ 1\ 0]_A$	$[0\ \ \ \eta_2\ 0]_A$
$[0\ 0\ 1]_A$	$[0\ \ 0\ \ \eta_3]_A$

and the matrix (A S A) can be written as

$$(A\ S\ A) = \begin{pmatrix} \eta_1 & 0 & 0 \\ 0 & \eta_2 & 0 \\ 0 & 0 & \eta_3 \end{pmatrix}$$

Each column of the deformation matrix represents the components of the new vector (referred to the original A basis) formed as a result of the deformation of a basis vector of A.

7

The strain represented by (A S A) is called a pure strain since its matrix representation (A S A) is symmetrical. This also means that it is possible to find three initially orthogonal directions (the principal axes) which remain orthogonal and unrotated during the deformation; a pure deformation consists of simple extensions or contractions along the principal axes. A vector parallel to a principal axis is not rotated by the deformation, but its magnitude may be altered. The ratio of its final length to initial length is the *principal deformation* associated with that principal axis. The directions $[1\ 0\ 0]_B$, $[0\ 1\ 0]_B$ and $[0\ 0\ 1]_B$ are therefore the principal axes of the Bain strain, and η_i the respective principal deformations. In the particular example under consideration, $\eta_1 = \eta_2$ so that any two perpendicular lines in the $(0\ 0\ 1)_B$ plane could form two of the principal axes. Since $\underline{a}_3 \| \underline{b}_3$ and since $\underline{a}_1$ and $\underline{a}_2$ lie in $(0\ 0\ 1)_B$, it is clear that the vectors $\underline{a}_i$ must also be the principal axes of the Bain deformation.

Since the deformation matrix (A S A) is referred to a basis system which coincides with the principal axes, the off-diagonal components of this matrix must be zero. Column (1) of the matrix (A S A) represents the new co-ordinates of the vector $[1\ 0\ 0]$, after the latter has undergone the deformation described by (A S A), and a similar rationale applies to the other columns. (A S A) deforms the FCC A lattice into a BCC α lattice, and the ratio of the final to initial volume of the material is simply $\eta_1\eta_2\eta_3$ (or more generally, the determinant of the deformation matrix). Finally, it should be noted that any tetragonality in the martensite can readily be taken into account by writing $\eta_3 = c/a_\gamma$, where c/a_α is the aspect ratio of the BCT martensite unit cell.

Example 2: The Bain Strain

Given that the principal distortions of the Bain strain (A S A), referred to the crystallographic axes of the FCC γ lattice (lattice parameter a_γ), are $\eta_1 = \eta_2 = 1.123883$, and $\eta_3 = 0.794705$, show that the vector

$$[A;\underline{x}] = [-0.645452\ \ 0.408391\ \ 0.645452]$$

remains undistorted, though not unrotated as a result of the operation of the Bain strain. Furthermore, show that for $\underline{x}$ to remain unextended as a result of the Bain strain, its components x_i must satisfy the equation

$$(\eta_1^2 - 1)x_1^2 + (\eta_2^2 - 1)x_2^2 + (\eta_3^2 - 1)x_3^2 = 0 \qquad \text{.....(6a)}$$

As a result of the deformation (A S A), the vector $\underline{x}$ becomes a new vector $\underline{y}$, according to the equation

$$(A\ S\ A)[A;\underline{x}] = [A;\underline{y}] = [\eta_1 x_1 \quad \eta_2 x_2 \quad \eta_3 x_3] = [-0.723412\ \ 0.458983\ \ 0.512944]$$

Now,

$$|\underline{x}|^2 = (\underline{x};A^*)[A;\underline{x}] = a_\gamma^2(x_1^2 + x_2^2 + x_3^2) \qquad \text{.....(6b)}$$

and,

$$|\underline{y}|^2 = (\underline{y};A^*)[A;\underline{y}] = a_\gamma^2(y_1^2 + y_2^2 + y_3^2) \qquad \text{.....(6c)}$$

Using these equations, and the numerical values of x_i and y_i obtained above, it is easy to show that $|\underline{x}| = |\underline{y}|$. It should be noted that although $\underline{x}$ remains unextended, it is rotated by the strain (A S A), since $x_i \neq y_i$. On equating (6b) to (6c) with $y_i = \eta_i x_i$, we get the required equation (6a). Since η_1 and η_2 are equal and greater than 1, and since η_3 is less than unity, eq.6a amounts to the equation of a right-circular cone, the axis of which coincides with $[0\ 0\ 1]_A$. Any vector initially lying on this cone will remain unextended as a result of the Bain Strain. This process can be illustrated by considering a spherical volume of the original austenite lattice; (A S A) deforms this into an ellipsoid of revolution, as illustrated in Fig. 5. Notice that the principal axes $(\underline{a}_i)$ remain unrotated by the deformation, and that lines such as ab and cd which become a'b' and c'd' respectively, remain unextended by the deformation (since they are all diameters of the original sphere), although rotated through the angle θ. The lines ab and cd of course lie on the initial cone described by eq.6a. Suppose now, that the ellipsoid resulting from the Bain strain is rotated through a right-handed angle of θ, about the axis a_2, then Fig. 5c illustrates that this rotation will cause the initial and final cones of unextended lines to touch along cd, bringing cd and c'd' into coincidence. If the total deformation can therefore be described as (A S A) combined with the above rigid body rotation, then such a deformation would leave the line cd both unrotated and unextended; such a deformation is called an *invariant line strain*. Notice that the total deformation, consisting of (A S A) and a rigid body rotation is no longer a pure strain, since the vectors parallel to the principal axes of (A S A) are rotated into the new positions a_i' (Fig. 5c).

It will later be shown that the lattice deformation in a martensitic transformation must contain an invariant line, so that the Bain strain must be combined with a suitable rigid body rotation in order to define the total lattice deformation. This explains why the experimentally observed orientation relationship (see Example 5) between martensite and austenite does not correspond to that implied by Fig. 1. The need to have an invariant line in the martensite-austenite interface means that the Bain Strain does not in itself constitute the total transformation strain, which can be factorised into the Bain Strain and a rigid body rotation. It is this total strain which determines the final orientation relationship although the Bain Strain accomplishes the total FCC to BCC lattice change. It is emphasised here that the Bain strain and the rotation are not physically distinct; the factorisation of the total transformation strain is simply a mathematical convenience.

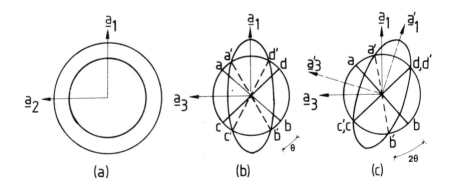

Fig. 5   (a) and (b) represent the effect of the Bain Strain on austenite, represented initially as a sphere of diameter ab which then deforms into an ellipsoid of revolution. (c) shows the invariant-line strain obtained by combining the Bain Strain with a rigid body rotation.

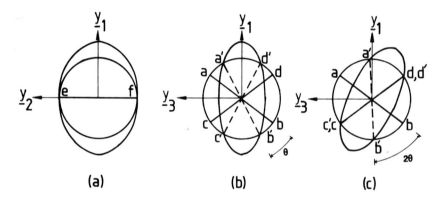

(a)   (b)   (c)

<u>Fig. 6</u>   Illustration of the strain (Y Q Y), the undeformed crystal represented initially as a sphere of diameter ef. (c) illustrates that a combination of (Y Q Y) with a rigid body rotation gives an invariant-plane strain.

Interfaces

A vector parallel to a principal axis of a pure deformation may become extended but is not changed in direction by the deformation. The ratio (η) of its final to initial length is called a principal deformation associated with that principal axis and the corresponding quantity ($\eta - 1$) is called a principal strain. Example 2 demonstrates that when two of the principal strains of the pure deformation differ in sign from the third, all three being non-zero, it is possible to obtain a total strain which leaves one line invariant. It intuitively seems advantageous to have the invariant-line in the interface connecting the two crystals, since their lattices would then match exactly along that line.

A completely undistorted interface would have to contain two non-parallel directions which are invariant to the total transformation strain. The following example illustrates the characteristics of such a transformation strain, called an invariant-plane strain, which allows the existence of a plane which remains unrotated and undistorted during the deformation.

Example 3: Deformations and Interfaces

A pure strain (Y Q Y), referred to an orthonormal basis Y whose basis vectors are parallel to the principal axes of the deformation, has the principal deformations $\eta_1 = 1.192281$, $\eta_2 = 1$ and $\eta_3 = 0.838728$. Show that (Y Q Y) combined with a rigid body rotation gives a total strain which leaves a plane unrotated and undistorted.

Because (Y Q Y) is a pure strain referred to a basis composed of unit vectors parallel to its principal axes, it consists of simple extensions or contractions along the basis vectors y_1, y_2 and y_3. Hence, Fig. 6 can be constructed as in the previous example. Since $\eta_2 = 1$, efIIy_2 remains unextended and unrotated by (Y Q Y), and if a rigid body rotation (about fe as the axis of rotation) is added to bring cd into coincidence with c'd', then the two vectors ef and ab remain invariant to the total deformation. Any combination of ef and ab will also remain invariant, and hence all lines in the plane containing ef and ab are invariant, giving an invariant plane. Thus, a pure strain when combined with a rigid body rotation can generate an invariant-plane strain if two of its principal strains have opposite signs, the third being zero. Since it is the pure strain which actually accomplishes the lattice change (the rigid body rotation causes no further lattice change), any two lattices related by a pure strain with these characteristics may be joined by a fully coherent interface.

(Y Q Y) actually represents the pure strain part of the total transformation strain required to change a FCC lattice to an HCP (hexagonal close packed) lattice, without any volume change, by shearing on the $(1\ 1\ 1)_\gamma$ plane, in the $[1\ 1\ \bar{2}]_\gamma$ direction, the magnitude of the shear being equal to half the twinning shear (see Chapter 3). Consistent with the proof given above, a fully coherent interface is observed experimentally when HCP martensite is formed in this manner.

11

ORIENTATION RELATIONSHIPS

A substantial part of research on polycrystalline materials is concerned with the accurate determination, assessment and theoretical prediction of orientation relationships between adjacent crystals. There are obvious practical applications, as in the study of anisotropy and texture and in various mechanical property assessments. The existence of a reproducible orientation relation between the parent and product phases might determine the ultimate morphology of any precipitates, by allowing the development of low interfacial energy facets. It is possible that vital clues about nucleation in the solid state might emerge from a detailed examination of orientation relationships, even though these can usually only be measured when the crystals concerned are well into the growth stage. Needless to say, the properties of interfaces depend critically on the relative dispositions of the crystals that they connect.

Perhaps the most interesting experimental observation is that the orientation relationships that are found to develop during phase transformations (or during deformation or recrystallisation experiments) are usually not random[6-8]. The frequency of occurrence of any particular orientation relation usually far exceeds the probability of obtaining it simply by taking the two separate crystals and joining them up in an arbitrary way.

This indicates that there are favoured orientation relations, perhaps because it is these which allow the best fit at the interface between the two crystals. This would in turn reduce the interface energy, and hence the activation energy for nucleation. Nuclei which, by chance, happen to be orientated in this manner would find it relatively easy to grow, giving rise to the non-random distribution mentioned earlier.

On the other hand, it could be the case that nuclei actually form by the homogeneous deformation of a small region of the parent lattice. The deformation, which transforms the parent structure to that of the product (e.g., Bain strain), would have to be of the kind which minimises strain energy. Of all the possible ways of accomplishing the lattice change by homogeneous deformation, only a few might satisfy the minimum strain energy criterion - this would again lead to the observed non-random distribution of orientation relationships. It is a major phase transformations problem to understand which of these factors really determine the existence of rational orientation relations. In this chapter we deal with the methods of adequately describing the relationships between crystals.

Cementite in Steels

Cementite (Fe_3C, referred to as θ) is probably the most common precipitate to be found in steels; it has a complex orthorhombic crystal structure and can precipitate from supersaturated ferrite or austenite. When it grows from ferrite, it usually adopts the Bagaryatski [9] orientation relationship (described in Example 4) and it is particularly interesting that precipitation can occur at temperatures below 400K in times too short to allow any substantial diffusion of iron atoms [10], although long range diffusion of carbon atoms is clearly necessary and indeed possible. It has therefore been suggested that the cementite lattice may sometimes be generated by the deformation of the ferrite lattice, combined with the necessary diffusion of carbon into the appropriate sites [10,11].

Shackleton and Kelly [12] showed that the plane of precipitation of cementite from ferrite is $\{1\ 0\ 1\}_\theta \| \{1\ 1\ 2\}_\alpha$. This is consistent with the habit plane containing the direction of maximum coherency between the θ and α lattices [10], i.e., $<1\ 0\ 1>_\theta \| <1\ 1\ \bar{1}>_\alpha$. Cementite formed during the tempering of martensite adopts many crystallographic variants of this habit plane in any given plate of martensite; in lower bainite it is usual to find just one such variant, with the habit plane inclined at some 60° to the plate axis. The reasons for this are not established, but the problem is discussed in detail in ref.13. Cementite which forms from austenite usually exhibits the Pitsch [14]

12

orientation relation with $[0\ 0\ 1]_\theta \| [\bar{2}\ 2\ 5]_\gamma$ and $[1\ 0\ 0]_\theta \| [5\ \bar{5}\ 4]_\gamma$ and a mechanism which involves the intermediate formation of ferrite has been postulated [10] to explain this orientation relationship.

Example 4: The Bagaryatski Orientation Relationship

Cementite (θ) has an orthorhombic crystal structure, with lattice parameters a = 4.5241, b = 5.0883 and c = 6.7416 Angstroms along the $[1\ 0\ 0]$, $[0\ 1\ 0]$ and $[0\ 0\ 1]$ directions respectively. When cementite precipitates from ferrite (α, BCC structure, lattice parameter a_α = 2.8662 Angstroms), the lattices are related by the Bagaryatski orientation relationship, given by:

$$[1\ 0\ 0]_\theta \| [0\ \bar{1}\ 1]_\alpha, \quad [0\ 1\ 0]_\theta \| [1\ \bar{1}\ \bar{1}]_\alpha, \quad [0\ 0\ 1]_\theta \| [2\ 1\ 1]_\alpha \qquad .....(7a)$$

(i) Derive the co-ordinate transformation matrix (α J θ) representing this orientation relationship, given that the basis vectors of θ and α define the orthorhombic and BCC unit cells of the cementite and ferrite, respectively.

(ii) Published stereograms of this orientation relation have the $(0\ \bar{2}\ 3)_\theta$ plane exactly parallel to the $(1\ 3\ 3)_\alpha$ plane. Show that this can only be correct if the ratio (b/c) = $8\sqrt{2}/15$.

The information given concerns just parallelisms between vectors in the two lattices. In order to find (α J θ), it is necessary to ensure that the magnitudes of the parallel vectors are also equal, since the magnitude must remain invariant to a co-ordinate transformation. If the constants k, g and m are defined as

$$k = |[1\ 0\ 0]_\theta| \ / \ |[0\ \bar{1}\ 1]_\alpha| \ = a/(a_\alpha\sqrt{2}) \ = \ 1.116120$$
$$g = |[0\ 1\ 0]_\theta| \ / \ |[1\ \bar{1}\ \bar{1}]_\alpha| \ = b/(a_\alpha\sqrt{3}) \ = \ 1.024957 \qquad .....(7b)$$
$$m = |[0\ 0\ 1]_\theta| \ / \ |[2\ 1\ 1]_\alpha| \ = c/(a_\alpha\sqrt{6}) \ = \ 0.960242$$

then multiplying $[0\ \bar{1}\ 1]_\alpha$ by k makes its magnitude equal to that of $[1\ 0\ 0]_\theta$; the constants g and m can similarly be used for the other two α vectors.

Recalling now our definition a co-ordinate transformation matrix, we note that each column of (α J θ) represents the components of a basis vector of θ in the α basis. For example, the first column of (α J θ) consists of the components of $[1\ 0\ 0]_\theta$ in the α basis, i.e., $[0\ \bar{k}\ k]$. It follows that we can derive (α J θ) simply by inspection of the relations 7a,b, so that

$$(\alpha\ J\ \theta) \quad = \quad \begin{pmatrix} 0.000000 & 1.024957 & 1.920485 \\ -1.116120 & -1.024957 & 0.960242 \\ 1.116120 & -1.024957 & 0.960242 \end{pmatrix}$$

The transformation matrix can therefore be deduced simply by inspection when the orientation relationship (7a) is stated in terms of the *basis vectors* of one of the crystals concerned (in this case, the basis vectors of θ are specified in 7a). On the other hand, orientation relationships can, and often are, specified in terms of vectors other then the basis vectors (see example 5). Also, electron diffraction patterns may not include direct information about basis vectors. A more general method of solving for $(\alpha \ J \ \theta)$ is presented below; this method is independent of the vectors used in specifying the orientation relationship:

From eq.2a and the relations 7a,b we see that

$$[0 \ \bar{k} \ k]_\alpha = (\alpha \ J \ \theta)[1 \ 0 \ 0]_\theta$$
$$[g \ \bar{g} \ \bar{g}]_\alpha = (\alpha \ J \ \theta)[0 \ 1 \ 0]_\theta \qquad .....(7c)$$
$$[2m \ m \ m]_\alpha = (\alpha \ J \ \theta)[0 \ 0 \ 1]_\theta$$

These equations can written as:

$$\begin{pmatrix} 0 & g & 2m \\ \bar{k} & \bar{g} & m \\ k & \bar{g} & m \end{pmatrix} = \begin{pmatrix} J_{11} & J_{12} & J_{13} \\ J_{21} & J_{22} & J_{23} \\ J_{31} & J_{32} & J_{33} \end{pmatrix} \begin{pmatrix} 1 & 0 & 0 \\ 0 & 1 & 0 \\ 0 & 0 & 1 \end{pmatrix} \qquad .....(7d)$$

where the J_{ij} (i = 1,2,3 & j = 1,2,3) are the elements of the matrix $(\alpha \ J \ \theta)$. From eq.7d, it follows that

$$(\alpha \ J \ \theta) = \begin{pmatrix} 0 & g & 2m \\ \bar{k} & \bar{g} & m \\ k & \bar{g} & m \end{pmatrix} = \begin{pmatrix} 0 & 1.024957 & 1.920485 \\ -1.116120 & -1.024957 & 0.960242 \\ 1.116120 & -1.024957 & 0.960242 \end{pmatrix}$$

It is easy to accumulate rounding off errors in such calculations and essential to use at least six figures after the decimal point.

To consider the relationships between *plane normals* (rather than directions) in the two lattices, we have to discover how vectors representing plane normals, (always referred to a reciprocal basis) transform. From eq.4, if $\underline{u}$ is any vector,

$$|\underline{u}|^2 = (\underline{u};\alpha^*)[\alpha;\underline{u}] = (\underline{u};\theta^*)[\theta;\underline{u}]$$

or

$$(\underline{u};\alpha^*)(\alpha \ J \ \theta)[\theta;\underline{u}] = (\underline{u};\theta^*)[\theta;\underline{u}]$$

giving

$$(\underline{u};\alpha^*)(\alpha \ J \ \theta) = (\underline{u};\theta^*)$$
$$(\underline{u};\alpha^*) = (\underline{u};\theta^*)(\theta \ J \ \alpha) \qquad .....(7e)$$

where

$$(\theta \ J \ \alpha) = (\alpha \ J \ \theta)^{-1}$$

$$(\theta \ J \ \alpha) = (1/(6gmk)) \begin{pmatrix} 0 & -3gm & 3gm \\ 2mk & -2mk & -2mk \\ 2gk & gk & gk \end{pmatrix} = \begin{pmatrix} 0 & -0.447981 & 0.447981 \\ 0.325217 & -0.325217 & -0.325217 \\ 0.347135 & 0.173567 & 0.173567 \end{pmatrix}$$

If $(\underline{u};\theta^*) = (0 \ \bar{2} \ 3)$ is now substituted into eq.7e, we get the corresponding vector

$$(\underline{u};\alpha^*) = (1/(6gmk))(6gk-4mk \quad 3gk+4mk \quad 3gk+4mk)$$

For this to be parallel to a (1 3 3) plane normal in the ferrite, the second and third components must equal three times the first; i.e.,

$$3(6lk - 4mk) = (3lk + 4mk), \text{ which is equivalent to } b/c = (8\sqrt{2})/15, \text{ as required.}$$

Finally, it should be noted that the determinant of $(\alpha \ J \ \theta)$ gives the ratio (volume of θ unit cell)/(volume of α unit cell). If the co-ordinate transformation simply involves a rotation of bases (e.g., when it describes the relation between two grains of identical structure), then the matrix is orthogonal and its determinant has a value of unity for all proper rotations (i.e., not involving

14

inversion operations). Such co-ordinate transformation matrices are called rotation matrices.

A stereographic representation of the Bagaryatski orientation is presented in Fig. 7. Stereograms are appealing because they provide a "picture" of the angular relationships between poles (plane normals) of crystal planes and give some indication of symmetry; the picture is of course distorted because distance on the stereogram does not scale directly with angle. Angular measurements on stereograms are in general made using Wulff nets and may typically be in error by a few degrees, depending on the size of the stereogram. Space and aesthetic considerations usually limit the number of poles plotted on stereograms, and those plotted usually have rational indices. Separate plots are needed for cases where directions and plane normals of the same indices have a different orientation in space. A co-ordinate transformation matrix is a precise way of representing orientation relationships; angles between *any* plane normals or directions can be calculated to any desired degree of accuracy and information on both plane normals and directions can be derived from just one matrix. With a little nurturing it is also possible to picture the meaning of the elements of a co-ordinate transformation matrix: each column of $(\alpha \ J \ \theta)$ represents the components of a basis vector of θ in the basis α, and the determinant of this matrix gives the ratio of volumes of the two unit cells.

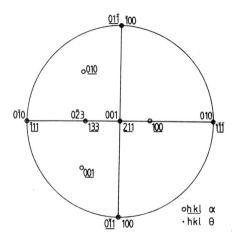

Fig. 7: Stereographic representation of the Bagaryatski orientation relationship between ferrite and cementite in steels, where

$$[1 \ 0 \ 0]_\theta \| [0 \ \bar{1} \ 1]_\alpha, \quad [0 \ 1 \ 0]_\theta \| [1 \ \bar{1} \ \bar{1}]_\alpha, \quad [0 \ 0 \ 1]_\theta \| [2 \ 1 \ 1]_\alpha$$

Note that these parallelisms are consistent with the co-ordinate transformation matrix $(\alpha \ J \ \theta)$ as derived in example 4:

$$(\alpha \ J \ \theta) = \begin{pmatrix} 0.000000 & 1.024957 & 1.920485 \\ -1.116120 & -1.024957 & 0.960242 \\ 1.116120 & -1.024957 & 0.0960242 \end{pmatrix}$$

Each column of $(\alpha \ J \ \theta)$ represents the components of a basis vector of θ in the basis of α.

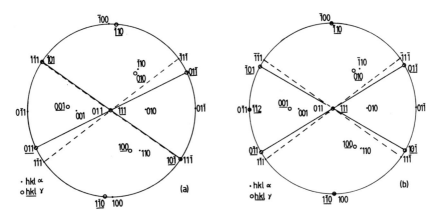

Fig. 8: (a) Stereographic representation of the Kurdjumov-Sachs orientation relationship. Note that the positions of the base vectors of the γ lattice are consistent with the matrix $(X \ J \ \gamma)$ derived in example 5:

$$(X \ J \ \gamma) = (a_\gamma / \ a_\alpha) \begin{pmatrix} 0.741582 & -0.666667 & -0.074915 \\ 0.649830 & 0.741582 & -0.166667 \\ 0.166667 & 0.074915 & 0.983163 \end{pmatrix}$$

Each column of $(X \ J \ \gamma)$ represents the components of a basis vector of γ in the basis X, so that $[1 \ 0 \ 0]_\gamma$, $[0 \ 1 \ 0]_\gamma$ and $[0 \ 0 \ 1]_\gamma$ are approximately parallel to $[1 \ 1 \ 0]_\alpha$, $[\bar{1} \ 1 \ 0]_\alpha$ and $[0 \ 0 \ 1]_\alpha$ respectively, as seen in the stereographic representation. (b) Stereographic representation of the Nishiyama-Wasserman orientation relationship. Note that this can be generated from the Kurdjumov-Sachs orientation by a rotation of 5.26^0 about $[011]_\alpha$. The necessary rotation makes $[\bar{1} \ \bar{1} \ 2]_\gamma$ exactly parallel to $[0 \ \bar{1} \ 1]_\alpha$ so that the Nishiyama-Wasserman orientation relation is also rational. In fact, the Nishiyama-Wasserman relation can be seen to be exactly midway between the two variants of the Kurdjumov-Sachs relation which share the same parallel close-packed plane. The stereograms also show that the Kurdjumov-Sachs and Nishiyama-Wasserman orientation relationships do not differ much from the γ/α orientation relationship implied by the Bain strain illustrated in Fig. 1.

Relations between FCC and BCC crystals

The ratio of the lattice parameters of austenite and ferrite in steels is about 1.24, and there are several other alloys (e.g., Cu-Zn brasses, Cu-Cr alloys) where FCC and BCC precipitates of similar lattice parameter ratios coexist. The orientation relations between these phases vary within a well defined range, but it is usually the case that a close-packed $\{1 \ 1 \ 1\}_{FCC}$ plane is approximately parallel to a $\{0 \ 1 \ 1\}_{BCC}$ plane (Fig. 8). Within these planes, there can be a significant variation in orientation, with $<\bar{1} \ 0 \ 1>_{FCC} || <\bar{1} \ \bar{1} \ 1>_{BCC}$ for the Kurdjumov-Sachs[15] orientation relation,

16

and $<\bar{1}\ 0\ 1>_{FCC}$ about 5.3° from $<\bar{1}\ \bar{1}\ 1>_{BCC}$ (towards $<1\ \bar{1}\ 1>_{BCC}$) for the Nishiyama-Wasserman relation[16]. It is experimentally very difficult to distinguish between these relations using ordinary electron diffraction or X-ray techniques, but very accurate work, such as that of Crosky et al. [17], clearly shows a spread of FCC-BCC orientation relationships (roughly near the Kurdjumov-Sachs and Nishiyama-Wasserman orientations) within the same alloy. Example 5 deals with the exact Kurdjumov-Sachs orientation relationship.

Example 5: The Kurdjumov-Sachs Orientation Relationship

Two plates of Widmanstatten ferrite (basis symbols X and Y respectively) growing in the same grain of austenite (basis symbol γ) are found to exhibit two different variants of the Kurdjumov-Sachs orientation relationship with the austenite; the data below shows the sets of parallel vectors of the three lattices:

$$[1\ 1\ 1]_{\gamma}\ \|\ [0\ 1\ 1]_X \qquad\qquad [1\ 1\ \bar{1}]_{\gamma}\ \|\ [0\ 1\ 1]_Y$$
$$[\bar{1}\ 0\ 1]_{\gamma}\ \|\ [\bar{1}\ \bar{1}\ 1]_X \qquad\qquad [1\ 0\ 1]_{\gamma}\ \|\ [1\ \bar{1}\ 1]_Y$$
$$[1\ \bar{2}\ 1]_{\gamma}\ \|\ [2\ \bar{1}\ 1]_X \qquad\qquad [1\ \bar{2}\ \bar{1}]_{\gamma}\ \|\ [2\ 1\ \bar{1}]_Y$$

Derive the matrices (X J γ) and (Y J γ). Hence obtain the rotation matrix (X J Y) describing the orientation relationship between the two Widmanstatten ferrite plates (the basis vectors of X, Y and γ define the respective conventional unit cells).

The information given relates to parallelisms between vectors in different lattices, and it is necessary to equalise the magnitudes of parallel vectors in order to solve for the various co-ordinate transformation matrices. Defining the constants k, g and m as

$$k = (a_{\gamma}\sqrt{3})/(a_{\alpha}\sqrt{2}) \qquad\qquad g = (a_{\gamma}\sqrt{2})/(a_{\alpha}\sqrt{3}) \qquad\qquad m = \sqrt{6}a_{\gamma}/(\sqrt{6}a_{\alpha}) = a_{\gamma}/\ a_{\alpha}$$

we obtain (as in eq.7c):

$$[\ 0\ k\ k]_X = (X\ J\ \gamma)[1\ 1\ 1]_{\gamma}$$
$$[\ \bar{g}\ \bar{g}\ g]_X = (X\ J\ \gamma)[\bar{1}\ 0\ 1]_{\gamma} \quad \text{or}$$
$$[2m\ \bar{m}\ m]_X = (X\ J\ \gamma)[1\ \bar{2}\ 1]_{\gamma}$$

$$\begin{pmatrix} 0 & \bar{g} & 2m \\ k & \bar{g} & \bar{m} \\ k & g & m \end{pmatrix} = \begin{pmatrix} J_{11} & J_{12} & J_{13} \\ J_{21} & J_{22} & J_{23} \\ J_{31} & J_{32} & J_{33} \end{pmatrix} \begin{pmatrix} 1 & \bar{1} & 1 \\ 1 & 0 & \bar{2} \\ 1 & 1 & 1 \end{pmatrix}$$

$$(X\ J\ \gamma) = \begin{pmatrix} 0 & \bar{g} & 2m \\ k & \bar{g} & \bar{m} \\ k & g & m \end{pmatrix} \begin{pmatrix} 2/6 & 2/6 & 2/6 \\ \bar{3}/6 & 0 & 3/6 \\ 1/6 & \bar{2}/6 & 1/6 \end{pmatrix} = (1/6) \begin{pmatrix} 3g+2m & \bar{4}m & \bar{3}g+2m \\ 2k+3g-m & 2k+2m & 2k-3g-m \\ 2k-3g+m & 2k-2m & 2k+3g+m \end{pmatrix}$$

17

so that

$$(X \ J \ \gamma) = (a_\gamma / a_\alpha) \begin{pmatrix} 0.741582 & -0.666667 & -0.074915 \\ 0.649830 & 0.741582 & -0.166667 \\ 0.166667 & 0.074915 & 0.983163 \end{pmatrix}$$

In an similar way, we find

$$(Y \ J \ \gamma) = (a_\gamma / a_\alpha) \begin{pmatrix} 0.741582 & -0.666667 & 0.074915 \\ 0.166667 & 0.074915 & -0.983163 \\ 0.649830 & 0.741582 & 0.166667 \end{pmatrix}$$

To find the rotation matrix relating X and Y, we proceed as follows:

$$[X;\underline{u}] = (X \ J \ \gamma)[\gamma;\underline{u}] \quad \text{and} \quad [Y;\underline{u}] = (Y \ J \ \gamma)[\gamma;\underline{u}] \quad \text{and} \quad [X;\underline{u}] = (X \ J \ Y)[Y;\underline{u}]$$

it follows that

$$(X \ J \ \gamma)[\gamma;\underline{u}] = (X \ J \ Y)[Y;\underline{u}]$$

substituting for $[Y;\underline{u}]$, we get

$$(X \ J \ \gamma)[\gamma;\underline{u}] = (X \ J \ Y)(Y \ J \ \gamma)[\gamma;\underline{u}]$$

so that

$$(X \ J \ Y) = (X \ J \ \gamma)(\gamma \ J \ Y)$$

carrying out this operation, we get the required X-Y orientation relation

$$(X \ J \ Y) = \begin{pmatrix} 0.988776 & 0.147308 & -0.024972 \\ -0.024972 & 0.327722 & 0.944445 \\ 0.147308 & -0.933219 & 0.327722 \end{pmatrix}$$

We see that the matrix (X J Y) is orthogonal because it represents an axis transformation between like orthogonal bases. In fact, (X J γ) and (Y J γ) each equal an orthogonal matrix times a scalar factor (a_γ/a_α); this is because the bases X, Y and γ are themselves orthogonal.

In the above example, we chose to represent the Kurdjumov-Sachs orientation relationship by a co-ordinate transformation matrix (X J γ). *Named* orientation relationships like this usually assume the parallelism of particular low index planes and directions and in the example under consideration, these parallelisms are independent of the lattice parameters of the FCC and BCC structures concerned. In such cases, the orientation relationship may be represented by a pure rotation matrix, relating the orthogonal, but not necessarily orthonormal, bases of the two structures. Orientation relationships are indeed often specified in this way, or in some equivalent manner such as an axis-angle pair or a set of three Euler angles. This provides an economic way of representing orientation relations, but it should be emphasized that there is a loss of information in doing this. For example, a co-ordinate transformation matrix like (X J γ) not only gives information about vectors which are parallel, but also gives a ratio of the volumes of the two unit cells.

Orientation Relationships between Grains of Identical Structure

The relationship between two crystals which are of identical structure but which are misorientated with respect to each other is described in terms of a rotation matrix representing the rigid body rotation which can be imagined to generate one crystal from the other. As discussed below, any rotation of this kind, which leaves the common origin of the two crystals fixed, can also be described in terms of a rotation of 180^0 or less about an axis passing through that origin.

Example 6: Axis-Angle Pairs, and Rotation Matrices

Two ferrite grains X and Y can be related by a rotation matrix

$$(Y \ J \ X) = (1/3) \begin{pmatrix} 2 & 2 & \bar{1} \\ \bar{1} & 2 & 2 \\ 2 & \bar{1} & 2 \end{pmatrix}$$

where the basis vectors of X and Y define the respective BCC unit cells. Show that the crystal Y can be generated from X by a right-handed rotation of 60° about an axis parallel to the $[1\ 1\ 1]_X$ direction.

A rigid body rotation leaves the magnitudes and relative positions of all vectors in that body unchanged. For example, an initial vector $\underline{u}$ with components $[u_1 \ u_2 \ u_3]_X$ relative to the X basis, due to rotation becomes a new vector $\underline{w}$, with the *same* components $[u_1 \ u_2 \ u_3]_Y$, but with respect to the rotated basis Y. The fact that $\underline{w}$ represents a different direction than $\underline{u}$ (due to the rotation operation) means that its components in the X basis , $[w_1 \ w_2 \ w_3]_X$ must differ from $[u_1 \ u_2 \ u_3]_X$. The components of $\underline{w}$ in either basis are obviously related by

$$[Y;\underline{w}] = (Y \ J \ X)[X;\underline{w}]$$

in other words, $[u_1 \ u_2 \ u_3] = (Y \ J \ X)[w_1 \ w_2 \ w_3]$ .....(8a)

However, if $\underline{u}$ happens to lie along the axis of rotation relating X and Y, then not only will $[X;\underline{u}] = [Y;\underline{w}]$ as before, but its *direction* also remains invariant to the rotation operation, so that $[X;\underline{w}] = [Y;\underline{w}]$. From eq.8a,

$$(Y \ J \ X)[X;\underline{w}] = [Y;\underline{w}]$$

so that $(Y \ J \ X)[X;\underline{u}] = [X;\underline{u}]$

and hence $\{(Y \ J \ X) - I\}[X;\underline{u}] = 0$ .....(8b)

where I is a 3x3 identity matrix. Any rotation axis must satisfy an equation of this form; expanding eq.8b, we get

$$-(1/3)u_1 + (2/3)u_2 - (1/3)u_3 = 0$$
$$-(1/3)u_1 - (1/3)u_2 + (2/3)u_3 = 0$$
$$(2/3)u_1 - (1/3)u_2 - (1/3)u_3 = 0$$

Solving these simultaneously gives $u_1 = u_2 = u_3$, proving that the required rotation axis lies along the $[1\ 1\ 1]_X$ direction, which is of course, also the $[1\ 1\ 1]_Y$ direction.

The angle, and sense of rotation can be determined by examining a vector $\underline{v}$ which lies at 90° to $\underline{u}$. If, say, $\underline{v} = [\bar{1}\ 0\ 1]_X$, then as a result of the rotation operation it becomes $\underline{z} = [\bar{1}\ 0\ 1]_Y = [0\ \bar{1}\ 1]_X$, making an angle of 60° with $\underline{v}$, giving the required angle of rotation. Since $\underline{v} \wedge \underline{z}$ gives $[1\ 1\ 1]_X$, it is also a rotation in the right-handed sense.

Comments

(i) The problem illustrates the fact that the orientation relation between two grains can be represented by a matrix such as (Y J X), or by an axis-angle pair such as $[1\ 1\ 1]_X$ and 60°. Obviously, the often used practice of stating a misorientation between grains in terms of just an angle of rotation is inadequate and incorrect.

(ii) If we always represent an axis of rotation as a unit vector (or in general, a vector of fixed magnitude), then only three independent quantities are needed to define a misorientation between grains: two components of the axis of rotation, and an angle of rotation. It follows that a rotation matrix must also have only three independent terms. In fact, the components of any rotation matrix can be written in terms of a vector $\underline{u} = [u_1\ u_2\ u_3]$ which lies along the axis of rotation (such that $u_1u_1+u_2u_2+u_3u_3= 1$), and in terms of the right-handed angle of rotation θ as follows:

$$(Y\ J\ X)\ =\ \begin{pmatrix} u_1u_1(1-m)+m & u_1u_2(1-m)+u_3n & u_1u_3(1-m)-u_2n \\ u_1u_2(1-m)-u_3n & u_2u_2(1-m)+m & u_2u_3(1-m)+u_1n \\ u_1u_3(1-m)+u_2n & u_2u_3(1-m)-u_1n & u_3u_3(1-m)+m \end{pmatrix} \quad .....(8c)$$

where $m = \cos(\theta)$ and $n = \sin(\theta)$

The right-handed angle of rotation can be obtained from the fact that

$$J_{11} + J_{22} + J_{33} = 1 + 2\cos(\theta) \quad .....(8d)$$

and the components of the vector $\underline{u}$ along the axis of rotation are given by

$$u_1 = (J_{23}-J_{32})/(2\sin(\theta))$$
$$u_2 = (J_{31}-J_{13})/(2\sin(\theta)) \quad .....(8e)$$
$$u_3 = (J_{12}-J_{21})/(2\sin(\theta))$$

From the definition of a co-ordinate transformation matrix, each column of $(Y\ J\ X)$ gives the components of a basis vector of X in the basis Y. It follows that

$[1\ 0\ 0]_X \| [2\ \bar{1}\ 2]_Y, \qquad\qquad [0\ 1\ 0]_X \| [2\ 2\ \bar{1}]_Y, \qquad\qquad [0\ 0\ 1]_X \| [\bar{1}\ 2\ 2\ ]$
Y

Suppose now that there exists another ferrite crystal (basis symbol Z), such that

$[0\ \bar{1}\ 0]_Z \| [2\ \bar{1}\ 2]_Y, \qquad\qquad [1\ 0\ 0]_Z \| [2\ 2\ \bar{1}]_Y, \qquad\qquad [0\ 0\ 1]_Z \| [\bar{1}\ 2\ 2]_Y$

$$\text{then}\quad (Y\ J\ Z)\ =\quad (1/3)\quad \begin{pmatrix} 2 & \bar{2} & \bar{1} \\ 2 & 1 & 2 \\ \bar{1} & \bar{2} & 2 \end{pmatrix}$$

with the crystal Y being generated from Z by a right-handed rotation of 70.52° about $[1\ 0\ \bar{1}]_Z$ direction. It can easily be demonstrated that

$$(Z\ J\ X)\ =\quad \begin{pmatrix} 0 & 1 & 0 \\ \bar{1} & 0 & 0 \\ 0 & 0 & 1 \end{pmatrix} \qquad\qquad \text{from } (Z\ J\ X) = (Z\ J\ Y)(Y\ J\ X)$$

so that crystal X can be generated from Z by a rotation of 90° about $[0\ 0\ 1]_X$ axis. However, this is clearly a symmetry operation of a cubic crystal, and it follows that crystal X can never be experimentally distinguished from crystal Z, so that the matrices $(Y\ J\ X)$ and $(Y\ J\ Z)$ are crystallographically equivalent, as are the corresponding axis-angle pair descriptions. In other words, Y can be generated from X either by a rotation of 60° about $[1\ 1\ 1]_X$, or by a rotation of 70.52° about $[\bar{1}\ 0\ 1]_X$. The two axis-angle pair representations are equivalent. There are actually 24 matrices like $(Z\ J\ X)$ which represent symmetry rotations in cubic systems. It follows that a cubic bicrystal can be represented in 24 equivalent ways, with 24 axis-angle pairs. Any rotation matrix like $(Y\ J\ X)$ when multiplied by rotation matrices representing symmetry operations (e.g., $(Z\ J\ X)$) will lead to the 24 axis-angle pair representations. The degeneracy of other structures is as follows[18]: Cubic (24), Hexagonal (12), Hexagonal close-packed (6), Tetragonal (8), Trigonal (6),

20

Orthorhombic (4), Monoclinic (2) and Triclinic (1). In general, the number N of axis-angle pairs is given by $N=1+N_2+2N_3+3N_4+5N_6$ where N_2, N_3, N_4 and N_6 refer to the number of diads, triads, tetrads and hexads in the symmetry elements of the structure concerned.

Fig. 9 below is an electron diffraction pattern taken from an internally twinned martensite plate in a Fe-4Ni-0.4C wt.pct steel. It contains two <0 1 1> BCC zones, one from the parent plate (m) and the other from the twin (t). The pattern clearly illustrates how symmetry makes it possible to represent the same bi-crystal in terms of more than one axis-angle pair. This particular pattern shows that the twin crystal can be generated from the parent in at least three different ways: (i) Rotation of 70.52^o about the <0 1 1> zone axes of the patterns, (ii) Rotation of 180° about the {1 1 $\bar{1}$} plane normal, and (iii) Rotation of 180° about the {2 $\bar{1}$ 1} plane normal. It is apparent that these three operations would lead the same orientation relation between the twin and the parent lattices.

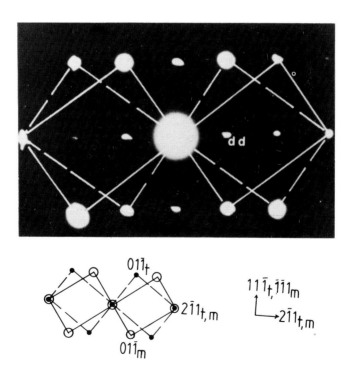

Fig. 9   Electron diffraction pattern from a martensite plate (m) and its twin (t). Spots not connected by lines (e.g., "dd") arise from double diffraction.

Example 7: "Double" Twinning

Plates of BCC martensite in Fe-30.4Ni wt.pct. contain $\{1\ 1\ \bar{2}\}$ transformation twins, the two twin orientations X and Y being related by a rotation of 60° about a $<1\ 1\ 1>$ axis. Deformation of this martensite at low temperatures leads to the formation of twins on $\{5\ 8\ 11\}$ planes, the direction of the twinning shear being $<\bar{5}\ \bar{1}\ 3>$. This is a very rare mode of twinning deformation in BCC systems; show that its occurrence may be related to the fact that such twins can propagate without any deviation, across the already existing transformation twins in the martensite [19,20].

The orientation relationship between the transformation twins is clearly the same as the matrix (Y J X) of Example 6. Using this matrix and eqs.2a,7e we obtain:

$$[\bar{5}\ \bar{1}\ 3]_X \| <\bar{5}\ 3\ \bar{1}>_Y$$
$$(5\ 8\ 11)_X \| \{5\ 11\ 8\}_Y$$

It follows that $\{5\ 8\ 11\}$ deformation twins can propagate without deviation across the transformation twins, since the above planes and directions, respectively, are crystallographically equivalent and indeed parallel. This may explain the occurrence of such an unusual deformation twinning mode.

The Metric

For cubic crystals, it is a familiar result that if the indices $[u_1\ u_2\ u_3]$ of a direction $\underline{u}$ in the lattice are numerically identical to the Miller indices $(h_1\ h_2\ h_3)$ of a plane in the lattice, then the normal to this plane $(\underline{h})$ is parallel to the direction mentioned. In other words, $\underline{u}\|\underline{h}$, and since $[A;\underline{u}] = [u_1\ u_2\ u_3]$ and $(\underline{h};A^*) = (h_1\ h_2\ h_3)$, we have $[u_1\ u_2\ u_3] = [h_1\ h_2\ h_3]$. (A represents the basis of the cubic system concerned, and A^* the corresponding reciprocal basis, in the usual way).

However, this is a special case reflecting the high symmetry of cubic systems, and it is not generally true that if $u_i = h_i$, then $\underline{u}\|\underline{h}$. For example, the $[1\ 2\ 3]$ direction in cementite is not parallel to the $(1\ 2\ 3)$ plane normal.

Consider any arbitrary crystal system, defined by a basis A (basis vectors $\underline{a}_i$), and by the corresponding reciprocal basis A^* consisting of the basis vectors $\underline{a}_i^*$ (obtained as in eq.3a). To find the angle between the direction $\underline{u}$ and the plane normal $\underline{h}$, it would be useful to have a matrix $(A^*\ G\ A)$, which allows the transformation of the components of a vector referred to the basis A, to those referred to its reciprocal basis A^*. (The symbol G is used, rather than J, to follow convention). This matrix, called the metric, with components G_{ij} can be determined in exactly the same manner as any other co-ordinate transformation matrix. Each column of $(A^*\ G\ A)$ thus consists of the components of one of the basis vectors of A, when referred to the basis A^*. For example,

$$\underline{a}_1 = G_{11}\underline{a}_1^* + G_{21}\underline{a}_2^* + G_{31}\underline{a}_3^* \qquad .....(9a)$$

Taking successive scalar dot products with $\underline{a}_1,\underline{a}_2$ and $\underline{a}_3$ respectively on both sides of eq.9a, we get

$$G_{11} = \underline{a}_1.\underline{a}_1,\ G_{21} = \underline{a}_1.\underline{a}_2,\ G_{31} = \underline{a}_1.\underline{a}_3$$

since $\underline{a}_i.\underline{a}_j^* = 0$ when $i \neq j$ (eq.4b). The rest of the elements of $(A^*\ G\ A)$ can be determined in a similar way, so that

$$(A^* \ G \ A) \quad = \quad \begin{pmatrix} \underline{a}_1 \cdot \underline{a}_1 & \underline{a}_2 \cdot \underline{a}_1 & \underline{a}_3 \cdot \underline{a}_1 \\ \underline{a}_1 \cdot \underline{a}_2 & \underline{a}_2 \cdot \underline{a}_2 & \underline{a}_3 \cdot \underline{a}_2 \\ \underline{a}_1 \cdot \underline{a}_3 & \underline{a}_2 \cdot \underline{a}_3 & \underline{a}_3 \cdot \underline{a}_3 \end{pmatrix} \qquad .....(9b)$$

It is easily demonstrated that the determinant of $(A^* \ G \ A)$ equals the square of the volume of the cell formed by the basis vectors of A. We note also that for orthonormal co-ordinates, $(Z^* \ G \ Z) = I$.

Example 8: Angles between plane normals and directions in an orthorhombic structure

A crystal with an orthorhombic structure has lattice parameters a, b and c. If the edges of the orthorhombic unit cell are taken to define the basis θ, determine the metric $(\theta^* \ G \ \theta)$. Hence derive the equation giving the angle ϕ between a plane normal $(\underline{h};\theta^*) = (h_1 \ h_2 \ h_3)$ and any direction $[\theta;\underline{u}] = [u_1 \ u_2 \ u_3]$.

From the definition of a scalar dot product, $(\underline{h}.\underline{u})/(|\underline{h}||\underline{u}|) = \cos\phi$. Now,

$$(\theta^* \ G \ \theta) = \begin{pmatrix} a^2 & 0 & 0 \\ 0 & b^2 & 0 \\ 0 & 0 & c^2 \end{pmatrix} \qquad (\theta \ G \ \theta^*) = \begin{pmatrix} 1/a^2 & 0 & 0 \\ 0 & 1/b^2 & 0 \\ 0 & 0 & 1/c^2 \end{pmatrix}$$

From eq.4,

$$\begin{aligned} |\underline{h}|^2 &= \underline{h}.\underline{h} = (\underline{h};\theta^*)[\theta;\underline{h}] \\ &= (\underline{h};\theta^*)(\theta \ G \ \theta^*)[\theta^*;\underline{h}] \\ &= h_1^2/a^2 + h_2^2/b^2 + h_3^2/c^2 \end{aligned}$$

Similarly,

$$\begin{aligned} |\underline{u}|^2 &= \underline{u}.\underline{u} = (\underline{u};\theta)[\theta^*;\underline{u}] \\ &= (\underline{u};\theta)(\theta^* \ G \ \theta)[\theta;\underline{u}] \\ &= u_1^2 a^2 + u_2^2 b^2 + u_3^2 c^2 \end{aligned}$$

Now, $\quad \underline{h}.\underline{u} = (\underline{h};\theta^*)[\theta;\underline{u}] = h_1 u_1 + h_2 u_2 + h_3 u_3$

so that $\cos\phi = \dfrac{(h_1 u_1 + h_2 u_2 + h_3 u_3)}{\{(h_1^2/a^2 + h_2^2/b^2 + h_3^2/c^2)(u_1^2 a^2 + u_2^2 b^2 + u_3^2 c^2)\}^{0.5}}$

More about the Vector Cross Product

Suppose that the basis vectors $\underline{a}$, $\underline{b}$ and $\underline{c}$ of the basis θ define an orthorhombic unit cell, then the cross product between two arbitrary vectors $\underline{u}$ and $\underline{v}$ referred to this basis may be written:

$$\underline{u} \wedge \underline{v} = (u_1\underline{a}+u_2\underline{b}+u_3\underline{c}) \wedge (v_1\underline{a}+v_2\underline{b}+v_3\underline{c})$$

where $[\theta;\underline{u}] = [u_1\ u_2\ u_3]$ and $[\theta;\underline{v}] = [v_1\ v_2\ v_3]$. This equation can be expanded to give:

$$\underline{u} \wedge \underline{v} = (u_2v_3-u_3v_2)(\underline{b} \wedge \underline{c}) + (u_3v_1-u_1v_3)(\underline{c} \wedge \underline{a}) + (u_2v_1-u_1v_2)(\underline{a} \wedge \underline{b})$$

Since $\underline{a}.\underline{b} \wedge \underline{c}=V$, the volume of the orthorhombic unit cell, and since $\underline{b} \wedge \underline{c}=V\underline{a}^*$(eq.3a), it follows that

$$\underline{u} \wedge \underline{v} = V\{(u_2v_3-u_3v_2)\underline{a}^* + (u_3v_1-u_1v_3)\underline{b}^* - (u_2v_1-u_1v_2)\underline{c}^*\} \qquad .....(10a)$$

Hence, $\underline{u} \wedge \underline{v}$ gives a vector whose components are expressed in the reciprocal basis. Writing $\underline{w} = \underline{u} \wedge \underline{v}$, with $(\underline{w};\theta^*) = (w_1\ w_2\ w_3)$, it follows that $w_1 = V(u_2v_3-u_3v_2)$, $w_2 = V(u_3v_1-u_1v_3)$ and $w_3 = V(u_2v_1-u_1v_2)$. The cross product of two directions thus gives a normal to the plane containing the two directions. If necessary, the components of $\underline{w}$ in the basis θ can easily be obtained using the metric, since $[\theta;\underline{w}] = (\theta\ G\ \theta^*)[\theta^*;\underline{w}]$. Similarly, the cross product of two vectors $\underline{h}$ and $\underline{k}$ which are referred to the reciprocal basis θ^*, such that $(\underline{h};\theta^*) = (h_1\ h_2\ h_3)$ and $(\underline{k};\theta^*) = (k_1\ k_2\ k_3)$, can be shown to be:

$$\underline{h} \wedge \underline{k} = (1/V)\{(h_2k_3-h_3k_2)\underline{a} + (h_3k_1-h_1k_3)\underline{b} - (h_2k_1-h_1k_2)\underline{c}\} \qquad .....(10b)$$

Hence, $\underline{h} \wedge \underline{k}$ gives a vector whose components are expressed in the real basis. The vector cross product of two plane normals gives a direction (zone axis) which is common to the two planes represented by the plane normals.

24

SLIP, TWINNING AND OTHER INVARIANT-PLANE STRAINS

The deformation of crystals by the conservative glide of dislocations on a single set of crystallographic planes causes shear in the direction of the resultant Burgers vector of the dislocations concerned, a direction which lies in the slip plane; the slip plane and slip direction constitute a slip system. The material in the slip plane remains crystalline during slip and since there is no reconstruction of this material during slip (e.g. by localised melting followed by resolidification), there can be no change in the relative positions of atoms in the slip plane; the atomic arrangement on the slip plane is thus completely unaffected by the deformation. Another mode of conservative plastic deformation is mechanical twinning, in which the parent lattice is homogeneously sheared into the twin orientation; the plane on which the twinning shear occurs is again unaffected by the deformation and can therefore form a coherent boundary between the parent and product crystals. If a material which has a Poisson's ratio equal to zero is uniaxially stressed below its elastic limit, then the plane that is normal to the stress axis is unaffected by the deformation since the only non-zero strain is that parallel to the stress axis (beryllium has a Poisson's ratio which is nearly zero).

All these strains belong to a class of deformations called *invariant-plane strains*. The operation of an invariant-plane strain (IPS) always leaves one plane of the parent crystal completely undistorted and unrotated; this plane is the invariant-plane. The condition for a strain to leave a plane undistorted is, as illustrated in example 3, that the principal deformations of its pure strain component, η_1, η_2 and η_3 are greater than, equal to and less than unity, respectively. However, as seen in Figs. 6a,b, this does not ensure that the undistorted plane is also unrotated; combination with a suitable rotation (Fig. 6c) produces the true invariant-plane. Before using the concept of an IPS to understand deformation and transformation theory, we develop a way of expressing invariant-plane strains which will considerably simplify the task [2].

In chapter 1, it was demonstrated that a homogeneous deformation (A S A) strains a vector $\underline{u}$ into another vector $\underline{v}$ which in general may have a different direction and magnitude:

$$[A;\underline{v}] = (A \ S \ A)[A;\underline{u}]. \qquad .....(11a)$$

However, the deformation could also have been defined with respect to another arbitrary basis, such as 'B' (basis vectors $\underline{b}_i$) to give the deformation matrix (B S B), with:

$$[B;\underline{v}] = (B \ S \ B)[B;\underline{u}]. \qquad .....(11b)$$

The physical effect of (B S B) on the vector $\underline{u}$ must of course be exactly the same as that of (A S A) on $\underline{u}$, and the initial and final vectors $\underline{u}$ and $\underline{v}$ remain unaffected by the change of basis (although their components change). If the co-ordinate transformation relating the bases A and B is given by (A J B), then:

$$[A;\underline{u}] = (A \ J \ B)[B;\underline{u}] \quad \text{and} \quad [A;\underline{v}] = (A \ J \ B)[B;\underline{v}].$$

Substituting these relations into eq.11a, we get

$$(A \ J \ B)[B;\underline{v}] = (A \ S \ A)(A \ J \ B)[B;\underline{u}]$$

or

$$[B;\underline{v}] = (B \ J \ A)(A \ S \ A)(A \ J \ B)[B;\underline{u}]$$

Comparison with eq.11b proves that

$$(B \ S \ B) = (B \ J \ A)(A \ S \ A)(A \ J \ B) \qquad .....(11c)$$

An equation like eq.11c represents a *Similarity Transformation*, changing the basis with respect to which the deformation is described, without altering the physical nature of the deformation.

We can now proceed to examine the nature of invariant-plane strains. Fig. 10 illustrates three such strains, defined with respect to a right-handed orthonormal basis Z, such that $\underline{z}_3$ is parallel to the unit normal $\underline{p}$ of the invariant-plane; $\underline{z}_1$ and $\underline{z}_2$ lie within the invariant-plane, $\underline{z}_1$ being parallel to the shear component of the strain concerned. Fig. 10a illustrates an invariant-plane strain which is purely dilatational, and is of the type to be expected when a plate-shaped precipitate grows diffusionally. The change of shape (as illustrated in Fig. 10a) due to the growth of this precipitate then reflects the volume change accompanying transformation.

In Fig. 10b, the invariant-plane strain corresponds to a simple shear, involving no change of volume, as in the homogeneous deformation of crystals by slip. The shape of the parent crystal alters in a way which reflects the shear character of the deformation.

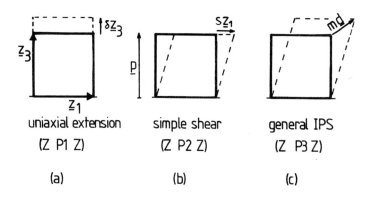

uniaxial extension   simple shear   general IPS

(Z P1 Z)   (Z P2 Z)   (Z P3 Z)

(a)   (b)   (c)

Fig. 10   Three kinds of Invariant-Plane Strains. The heavy lines indicate the shape before deformation. δ, s and m represent the magnitudes of the dilatational strain, shear strain and general displacement respectively.

The most general invariant-plane strain (Fig. 10c) involves both a volume change and a shear; if $\underline{d}$ is a unit vector in the direction of the displacements involved, then $m\underline{d}$ represents the displacement vector, where m is a scalar giving the magnitude of the displacements. $m\underline{d}$ may be factorised as $m\underline{d}$ = $s\underline{z}_1 + \delta\underline{z}_3$, where s and δ are the shear and dilatational components, respectively, of the invariant-plane strain. The strain illustrated in Fig. 10c is of the type associated with the martensitic transformation of γ iron into HCP iron. This involves a shear on the $\{1\ 1\ 1\}_\gamma$ planes in $<1\ 1\ \bar{2}>_\gamma$ direction, the magnitude of the shear being $8^{-1/2}$. However, there is also a dilatational component to the strain, since HCP iron is more dense than FCC iron (consistent with the fact that HCP iron is the stable form at high pressures); there is therefore a volume contraction on martensitic transformation, an additional displacement δ normal to the $\{1\ 1\ 1\}$ austenite planes.

It has often been suggested that the passage of a single Shockley partial dislocation on a close-packed plane of austenite leads to the formation of a 3-layer thick region of HCP, since this region contains the correct stacking sequence of close-packed planes for the HCP lattice. Until recently it has not been possible to prove this, because such a small region of HCP material gives very diffuse and unconvincing HCP reflections in electron diffraction experiments. However, the δ component of the FCC-HCP martensite transformation strain has recently been detected [21] to be present for single stacking faults, proving the HCP model of such faults.

Turning now to the description of the strains illustrated in Fig. 10, we follow the procedure of Chapter 1, to find the matrices (Z P Z); the symbol P in the matrix representation is used to identify specifically, an invariant-plane strain, the symbol S being the representation of any general deformation. Each column of such a matrix represents the components of a new vector generated by deformation of a vector equal to one of the basis vectors of Z. It follows that the three matrices representing the deformations of Fig. 10a-c are, respectively,

$$(Z\ P1\ Z)=\begin{pmatrix}1&0&0\\0&1&0\\0&0&1+\delta\end{pmatrix}\quad (Z\ P2\ Z)=\begin{pmatrix}1&0&s\\0&1&0\\0&0&1\end{pmatrix}\quad (Z\ P3\ Z)=\begin{pmatrix}1&0&s\\0&1&0\\0&0&1+\delta\end{pmatrix}$$

These matrices have a particularly simple form, because the basis Z has been chosen carefully, such that $\underline{p}\|\underline{z}_3$ and the direction of the shear is parallel to $\underline{z}_1$. However, it is often necessary to represent invariant-plane strains in a crystallographic basis, or in some other basis X. This can be achieved with the similarity transformation law, eq.11c. If (X J Z) represents the co-ordinate transformation from the basis Z to X, we have

$$(X\ P\ X) = (X\ J\ Z)(Z\ P\ Z)(Z\ J\ X)$$

Expansion of this equation gives[22]

$$(X\ P\ X) = \begin{pmatrix}1+md_1p_1&md_1p_2&md_1p_3\\md_2p_1&1+md_2p_2&md_2p_3\\md_3p_1&md_3p_2&1+md_3p_3\end{pmatrix}\quad .....(11d)$$

where d_i are the components of $\underline{d}$ in the X basis, such that $(\underline{d};X^*)[X;\underline{d}] = 1$. The vector $\underline{d}$ points in the direction of the displacements involved; a vector which is parallel to $\underline{d}$ remains parallel following deformation, although the ratio of its final to initial length may be changed. The quantitites p_i are the components of the invariant-plane normal $\underline{p}$, referred to the X^* basis, normalised to satisfy $(\underline{p};X^*)[X;\underline{p}] = 1$.

Eq.11d may be simplified as follows:

$$(X\ P\ X) = I + m[X;\underline{d}](\underline{p};X^*).\quad .....(11e)$$

The multiplication of a single-column matrix with a single-row matrix gives a 3x3 matrix, whereas the reverse order of multiplication gives a scalar quantity. The matrix (X P X) can be used to study the way in which vectors representing directions (referred to the X basis) deform. In order to examine the way in which vectors which are plane normals (i.e., referred to the reciprocal basis X^*) deform, we proceed in the following manner.

The property of a homogeneous deformation is that points which are originally colinear remain colinear after the deformation [5]. Also, lines which are initially coplanar remain coplanar following deformation. It follows that an initial direction $\underline{u}$ which lies in a plane whose normal is initially $\underline{h}$, becomes a new vector $\underline{v}$ within a plane whose normal is $\underline{k}$, where $\underline{v}$ and $\underline{k}$ result from the deformation of $\underline{u}$ and $\underline{h}$, respectively. Now, $\underline{h}.\underline{u} = \underline{k}.\underline{v} = 0$, so that:

$$(\underline{h};X^*)[X;\underline{u}] = (\underline{k};X^*)[X;\underline{v}] = (\underline{k};X^*)(X\ P\ X)[X;\underline{u}]$$

i.e.,
$$(\underline{k};X^*) = (\underline{h};X^*)(X\ P\ X)^{-1}\quad .....(12)$$

Eq.12 describes the way in which plane normals are affected by the deformation (X P X). From Eq.11e, it can be shown that

$$(X\ P\ X)^{-1} = I - mq[X;\underline{d}](\underline{p};X^*)\quad .....(13)$$

where $1/q = \det(X\ P\ X) = 1+m(\underline{p};X^*)[\underline{d};X]$. The inverse of (X P X) is thus another invariant-plane strain in the opposite direction.

Example 11: Tensile tests on single-crystals

A thin cylindrical single-crystal specimen of α iron is tensile tested at -140°C, the tensile axis being along the [4 4 1] direction (the cylinder axis). On application of a tensile stress, the entire specimen deforms by twinning on the (1 1 2) plane and in the [1 1 $\bar{1}$] direction, the magnitude of the twinning shear being $2^{-1/2}$. Calculate the plastic strain recorded along the tensile axis, assuming that the ends of the specimen are always maintained in perfect alignment. (Refs.23-26 contain details on single crystal deformation).

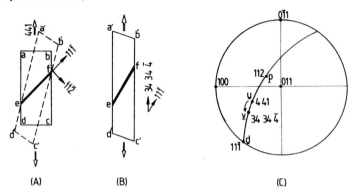

(A)   (B)   (C)

Fig. 11   Longitudinal section of the tensile specimen illustrating the (1 $\bar{1}$ 0) plane. All directions refer to the parent crystal basis. The tensile axis rotates towards $\underline{d}$, in the plane containing the original direction of the tensile axis ($\underline{u}$) and $\underline{d}$.

Fig. 11a illustrates the deformation involved; the parent crystal basis α consists of basis vectors which form the conventional BCC unit cell of α-iron. The effect of the mechanical twinning is to alter the original shape abcd to a'b'c'd'. ef is a trace of (1 1 2)$_\alpha$ on which the twinning shear occurs in the [1 1 $\bar{1}$]$_\alpha$ direction. However, as in most ordinary tensile tests, the ends of the specimen are constrained to be vertically aligned at all times; a'd' must therefore be vertical and the deforming crystal must rotate to comply with this requirement. The configuration illustrated in Fig. 11c is thus obtained, with ad and a'd' parallel, the tensile strain being (a'd'-ad)/(ad).

As discussed earlier, mechanical twinning is an invariant-plane strain; it involves a homogeneous simple shear on the twinning plane, a plane which is not affected by the deformation and which is common to both the parent and twin regions. Eq.11d can be used to find the matrix (α P α) describing the mechanical twinning, given that the normal to the invariant-plane is $(\underline{p};\alpha^*) = a_\alpha(6^{-1/2})(1\ 1\ 2)$, the displacement direction is $[\alpha;\underline{d}] = a_\alpha^{-1}(3^{-1/2})[1\ 1\ \bar{1}]$ and $m = 2^{-1/2}$. It should be noted that $\underline{p}$ and $\underline{d}$ respectively satisfy the conditions $(\underline{p};\alpha^*)[\alpha;\underline{p}] = 1$ and $(\underline{d};\alpha^*)[\alpha;\underline{d}] = 1$, as required for eq.11d. Hence

$$(\alpha\ P\ \alpha)\ =\ (1/6)\begin{pmatrix} 7 & 1 & 2 \\ 1 & 7 & 2 \\ \bar{1} & \bar{1} & 4 \end{pmatrix}$$

Using this, we see that an initial vector $[\alpha;\underline{u}] = [4\ 4\ 1]$ becomes a new vector $[\alpha;\underline{v}] = (\alpha\ P\ \alpha)[\alpha;\underline{u}] = (1/6)[34\ 34\ \bar{4}]$ due to the deformation. The need to maintain the specimen ends in alignment means that $\underline{v}$ is rotated to be parallel to the tensile axis. Now, $|\underline{u}| = 5.745a_\alpha$ where a_α is the lattice parameter of the ferrite, and $|\underline{v}| = 8.042a_\alpha$, giving the required tensile strain as $(8.042-5.745)/5.745 = 0.40$.

28

Comments

(i) From Fig. 11 it is evident that the end faces of the specimen will also undergo deformation (ab to a"b") and if the specimen gripping mechanism imposes constraints on these ends, then the rod will tend to bend into the form of an 'S'. For thin specimens this effect may be small.

(ii) The tensile axis at the beginning of the experiment was [4 4 1], but at the end is $(1/6)[34\ 34\ \overline{4}]$. The tensile direction has clearly rotated during the course of the experiment. The direction in which it has moved is $(1/6)[34\ 34\ \overline{4}]$ - $[4\ 4\ 1]$ = $(1/6)[10\ 10\ \overline{1}]$, parallel to $[1\ 1\ \overline{1}]$, the shear direction $\underline{d}$. In fact, any initial vector $\underline{u}$ will be displaced towards $\underline{d}$ to give a new vector $\underline{v}$ as a consequence of the IPS. Using eq.11e, we see that

$$[\alpha;\underline{v}] = (\alpha\ P\ \alpha)[\alpha;\underline{u}] = [\alpha;\underline{u}] + m[\alpha;\underline{d}](\underline{p};\alpha^*)[\alpha;\underline{u}] = [\alpha;\underline{u}] + \beta[\alpha;\underline{d}]$$

where β is a scalar quantity $\beta = m(\underline{p};\alpha^*)[\alpha;\underline{u}]$

Clearly, $\underline{v} = \underline{u} + \beta\underline{d}$, with $\beta = 0$ if $\underline{u}$ lies in the invariant-plane. All points in the lattice are thus displaced in the direction $\underline{d}$, although the extent of displacement depends on β.

(iii) Suppose now that only a volume fraction V of the specimen underwent the twinning deformation, the remainder being unaffected by the applied stress. The tensile strain recorded over the whole specimen as the gauge length would simply be 0.40V, which is obtained [24] by replacing m in eq.11d by Vm.

(iv) If the shear strain is allowed to vary, as in normal slip deformation, then the position of the tensile axis is still given by $\underline{v} = \underline{u} + \beta\underline{d}$, with β and $\underline{v}$ both varying as the test progresses. Since $\underline{v}$ is a linear combination of $\underline{u}$ and $\beta\underline{d}$, it must always lie in the plane containing both $\underline{u}$ and $\underline{d}$. Hence, the tensile axis rotates in the direction $\underline{d}$ within the plane defined by the original tensile axis and the shear direction, as illustrated in Fig. 11c.

Considering further the deformation of single-crystals, an applied stress σ can be resolved into a shear stress τ acting on a slip system. The relationship between σ and τ can be shown [23-25] to be $\tau = \sigma\ \cos\phi\ \cos\lambda$, where ϕ is the angle between the slip plane normal and the tensile axis, and λ is the angle between the slip direction and the tensile axis. Glide will first occur in the particular slip system for which τ exceeds the critical resolved shear stress necessary to initiate dislocation motion on that system. In austenite, glide is easiest on $\{1\ 1\ 1\}<0\ 1\ \overline{1}>$ and the γ standard projection (Fig. 12a) can be used [23] to determine the particular slip system which has the maximum resolved shear stress due to a tensile stress applied along $\underline{u}$. For example, if $\underline{u}$ falls within the stereographic triangle labelled A2, then $(\overline{1}\ 1\ 1)[0\ \overline{1}\ 1]$ can be shown to be the most highly stressed system. Hence, when τ reaches a critical value (the critical resolved shear stress), this system alone operates, giving "easy glide" since there is very little work hardening at this stage of deformation; the dislocations which accomplish the shear can simply glide out of the crystal and there is no interference with this glide since none of the other slip systems are active. Of course, the tensile axis is continually rotating towards $\underline{d}$ and may eventually fall on the boundary between two adjacent triangles in Fig. 12a. If $\underline{u}$ falls on the boundary between triangles A2 and D4, then the slip systems $(\overline{1}\ 1\ 1)[0\ \overline{1}\ 1]$ and $(1\ \overline{1}\ 1)[\overline{1}\ 0\ 1]$ are both equally stressed. This means that both systems can simultaneously operate and *duplex slip* is said to occur; the work hardening rate drastically increases as dislocations moving on different planes interfere with each other in a way which hinders glide and increases the defect density. It follows that a crystal which is initially orientated for single slip eventually deforms by multiple slip.

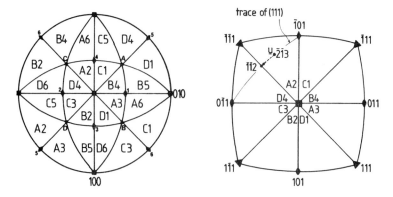

Stereographic analysis of slip in FCC single-crystals

Example 12: The Transition from Easy Glide to Duplex Slip

A single-crystal of austenite is tensile tested at 25°C, the stress being applied along $[\bar{2}\ \bar{1}\ 3]$ direction; the specimen deforms by easy glide on the $(\bar{1}\ 1\ 1)[0\ \bar{1}\ 1]$ system. If slip can only occur on systems of this form, calculate the tensile strain necessary for the onset of duplex slip. Assume that the ends of the specimen are maintained in alignment throughout the test.

The tensile axis ($\underline{v}$) is expected to rotate towards the slip direction, its motion being confined to the plane containing the initial tensile axis ($\underline{u}$) and the slip direction ($\underline{d}$). In Fig. 12b, $\underline{v}$ will therefore move on the trace of the $(1\ 1\ 1)$ plane. Duplex slip is expected to begin when $\underline{v}$ reaches the great circle which separates the stereographic triangles A2 and D4 of Fig. 12b, since the $(1\ \bar{1}\ 1)[\bar{1}\ 0\ 1]$ slip system will have a resolved shear stress equal to that on the initial slip system. The tensile axis can be expressed as a function of the shear strain m as in example 11:

$$[\gamma;\underline{v}] = [\gamma;\underline{u}] + m[\gamma;\underline{u}](\underline{p};\gamma^*)[\gamma;\underline{u}]$$

where $(\underline{p};\gamma^*) = (3^{-1/2})(\bar{1}\ 1\ 1)$ and $[\gamma;\underline{d}] = (2^{-1/2})[0\ \bar{1}\ 1]$, so that

$$[\gamma;\underline{v}] = [\gamma;\underline{u}] + (4m/6^{-1/2})[0\ \bar{1}\ 1] = [\bar{2}\ \bar{1}\ 3] + (4m/6^{-1/2})[0\ \bar{1}\ 1]$$

When duplex slip occurs, $\underline{v}$ must lie along the intersection of the $(1\ 1\ 1)$ and $(\bar{1}\ 1\ 0)$ planes, the former being the plane on which $\underline{v}$ is confined to move and the latter being the boundary between triangles A2 and D4. It follows that $\underline{v}\|[1\ 1\ \bar{2}]$ and must be of the form $\underline{v} = [v\ v\ 2\bar{v}]$. Substituting this into the earlier equation gives

$$[v\ v\ 2\bar{v}] = [\bar{2}\ 1\ 3] + (4m/6^{-1/2})[0\ \bar{1}\ 1]$$

and on comparing coefficients from both sides of this equation, we obtain

$$[\gamma;\underline{v}] = [\bar{2}\ \bar{2}\ 4]$$

so that the tensile strain required is $(|\underline{v}|-|\underline{u}|)/|\underline{u}| = 0.31$.

Deformation Twins

We can now proceed to study twinning deformations [4,25,27] in greater depth, noting that a twin is said to be any region of a parent which has undergone a homogeneous shear to give a re-orientated region with the same crystal structure. The example below illustrates some of the important concepts of twinning deformation.

Example 13: Twins in FCC crystals

Show that the austenite lattice can be twinned by a shear deformation on the $\{1\ 1\ 1\}$ plane and in the $<1\ 1\ \bar{2}>$ direction. Deduce the magnitude of the twinning shear, and explain why this is the most common twinning mode in FCC crystals. Derive the matrix representing the orientation relationship between the twin and parent lattices.

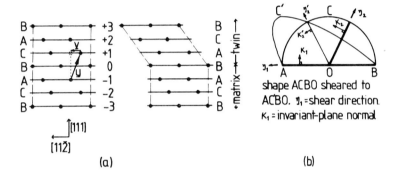

Fig. 13 Twinning in the FCC austenite lattice. The diagrams represent sections in the $(1\ \bar{1}\ 0)$ plane. In Fig. 13b, K'_2 and η'_2 are the final positions of the undistorted plane K_2 and the undistorted direction η_2, respectively.

$\{1\ 1\ 1\}$ planes in FCC crystals are close-packed, with a stacking sequence ..ABCABCABC... The region of the parent which becomes re-orientated due to the twinning shear can be generated by reflection across the twinning plane; the stacking sequence across the plane <u>B</u> which is the coherent twin interface is therefore ...ABCA<u>B</u>ACBA... Fig. 13a illustrates how a stack of close-packed planes (stacking sequence ..ABC..) may be labelled ..-1,0,+1... Reflection across 0 can be achieved by shearing atoms in the +1 plane into positions which are directly above (i.e., along $<1\ 1\ 1>$) the atoms in the -1 plane.

Fig. 13a is a section of the lattice on the $(\bar{1}\ 1\ 0)$ plane; it is evident that a displacement of all the atoms on +1 through a distance $v = |v|$ along $<1\ 1\ \bar{2}>$ gives the required reflection across the twinning plane 0. The twinning shear s is given by the equation $s^2 = (v/d)^2$, where d is the spacing of the $(1\ 1\ 1)$ planes. Since $v^2 = u^2 - 4d^2$, we may write

$$s^2 = (u/d)^2 - 4 \qquad .....(14)$$

where $u = |u|$ and $\underline{u}$ connects a site on the +1 plane to an equivalent site on the -1 plane (Fig. 13a). Hence, the FCC lattice can be twinned by a shear of magnitude $s = 2^{-1/2}$ on $\{1\ 1\ 1\}$.

To answer why a crystal twins in a particular way, it is necessary to make the physically reasonable assumption that the twinning mode should lead to the smallest possible shear (s). When the twin is forced to form in a constrained environment (as within a polycrystalline material), the shape change resulting from the shear deformation causes elastic distortions in both the twin and the matrix. The consequent strain energy increase (per unit volume of material) is approximately given by [28-30] $E = (c/r)\mu s^2$, where c and r represent the thickness and length of the twin respectively, and μ is the shear modulus. This is also the reason why mechanical twins have a lenticular morphology, since the small thickness to length ratio of thin-plates minimises E. Annealing twins grow diffusionally and there is no physical deformation involved in their growth. Hence, their shape is not restricted to that of a thin plate, the morphology being governed by the need to minimise interface energy. It is interesting that annealing and mechanical twins are crystallographically equivalent (if we ignore the absence of a shape change in the former) but their mechanisms of growth are very different.

Eq.14 indicates that s can be minimised by choosing twinning planes with large d spacings and by choosing the smallest vector $\underline{u}$ connecting a site on the +1 plane to an equivalent site on the -1 plane; for the (1 1 1) plane the smallest $\underline{u}$ is 0.5[1 1 2], as illustrated in Fig. 13a. Eq.14 can also be used to show that none of the planes of slightly smaller spacing than {1 1 1} can lead to twins with $s<2^{-1/2}$; two of these planes are also mirror planes and thus cannot serve as the invariant-plane (K_1, Fig. 13b) of the reflection twin.

From Fig. 13a we see that the twin lattice could also have been obtained by displacing the atoms in the +1 plane through a distance 2v along $[\bar{1}\ \bar{1}\ 2]$ had $\underline{u}$ been chosen to equal $[2^{1/2}\ 2^{1/2}\ 0]$, giving $s = 2^{1/2}$. This larger shear is of course inconsistent with the hypothesis that the favoured twinning mode involves the smallest shear, and indeed, this mode of twinning is not observed. To obtain the smallest shear, the magnitude of the vector $\underline{v}$ must also be minimised; in the example under consideration, the correct choice of $\underline{v}$ will connect a lattice site of plane +1 with the projection of its nearest neighbour lattice site on plane -1. The twinning direction is therefore expected to be along $[1\ 1\ \bar{2}]$. It follows that the operative twin mode for the FCC lattice should involve a shear of magnitude $s=2^{-1/2}$ on $\{1\ 1\ 1\}<1\ 1\ \bar{2}>$.

The matrix-twin orientation relationship (M J T) can be deduced from the fact that the twin was generated by a shear which brought atoms in the twin into positions which are related to the parent lattice points by reflection across the twinning plane (the basis vectors of M and T define the FCC unit cells of the matrix and twin crystals respectively). From Fig. 13 we note that:

$[1\ 1\ \bar{2}]_M \| [1\ 1\ \bar{2}]_T$ $\qquad\qquad$ $[\bar{1}\ 1\ 0]_M \| [\bar{1}\ 1\ 0]_T$ $\qquad\qquad$ $[1\ 1\ 1]_M \| [\bar{1}\ \bar{1}\ \bar{1}]_T$

It follows that

$$\begin{pmatrix} 1 & \bar{1} & 1 \\ 1 & 1 & 1 \\ \bar{2} & 0 & 1 \end{pmatrix} = \begin{pmatrix} J_{11} & J_{12} & J_{13} \\ J_{21} & J_{22} & J_{23} \\ J_{31} & J_{32} & J_{33} \end{pmatrix} \begin{pmatrix} 1 & \bar{1} & \bar{1} \\ 1 & 1 & \bar{1} \\ \bar{2} & 0 & \bar{1} \end{pmatrix}$$

Solving for (M J T), we get

$$(M\ J\ T) = (1/6) \begin{pmatrix} 1 & \bar{1} & 1 \\ 1 & 1 & 1 \\ \bar{2} & 0 & 1 \end{pmatrix} \begin{pmatrix} 1 & 1 & \bar{2} \\ \bar{3} & 3 & 0 \\ \bar{2} & \bar{2} & \bar{2} \end{pmatrix} = (1/3) \begin{pmatrix} 1 & \bar{2} & \bar{2} \\ \bar{2} & 1 & \bar{2} \\ \bar{2} & \bar{2} & 1 \end{pmatrix}$$

Comments

(i) Equations like eq.14 can be used [4] to predict the likely ways in which different lattices might

32

twin, especially when the determining factor is the magnitude of the twinning shear.

(ii) There are actually four different ways of generating the twin lattice from the parent crystal: (a) by reflection about the K_1 plane on which the twinning shear occurs, (b) by a rotation of π about η_1, the direction of the twinning shear, (c) by reflection about the plane normal to η_1 and (d) by a rotation of π about the normal to the K_1 plane.

Since most metals are centrosymmetric, operations (a) and (d) produce crystallographically equivalent results, as do (b) and (c). In the case of the FCC twin discussed above, the high symmetry of the cubic lattice means that all four operations are crystallographically equivalent. Twins which can be produced by the operations (a) and (d) are called type I twins; type II twins result form the other two twinning operations. The twin discussed in the above example is called a compound twin, since type I and type II twins cannot be crystallographically distinguished. Fig. 13b illustrates some additional features of twinning. The K_2 plane is the plane which (like K_1) is undistorted by the twinning shear, but unlike K_1, is rotated by the shear. The "plane of shear" is the plane containing η_1 and the perpendicular to K_1; its intersection with K_2 defines the undistorted but rotated direction η_2. In general, η_2 and K_1 are rational for type I twins, and η_1 and K_2 are rational for type II twins. The set of four twinning elements K_1, K_2, η_1 and η_2 are all rational for compound twins. From Fig. 13b, η_2 makes an angle of $\arctan(s/2)$ with the normal to K_1 and simple geometry shows that $\eta_2 = [1\ 1\ 2]$ for the FCC twin of example 13. The corresponding K_2 plane which contains η_2 and $\eta_1 \wedge \eta_2$ is therefore $(1\ 1\ \bar{1})$, giving the rational set of twinning elements

$$K_1 = (1\ 1\ 1) \quad \eta_2 = [1\ 1\ 2] \quad s = 2^{-1/2} \quad \eta_1 = [1\ 1\ \bar{2}] \quad K_2 = (1\ 1\ \bar{1})$$

In fact, it is only necessary to specify either K_1 and η_2 or K_2 and η_1 to completely describe the twin mode concerned.

The deformation matrix (M P M) describing the twinning shear can be deduced using eq.11d and the information $[M;\underline{d}]\|[1\ 1\ \bar{2}]$, $(\underline{p};M^*)\|(1\ 1\ 1)$ and $s = 2^{-1/2}$ to give

$$(M\ P\ M) = (1/6)\begin{pmatrix} 7 & 1 & 1 \\ 1 & 7 & 1 \\ \bar{2} & \bar{2} & 4 \end{pmatrix} \text{ and } (M\ P\ M)^{-1} = (1/6)\begin{pmatrix} 5 & \bar{1} & \bar{1} \\ \bar{1} & 5 & \bar{1} \\ 2 & 2 & 8 \end{pmatrix} \quad .....(15a)$$

and if a vector $\underline{u}$ is deformed into a new vector $\underline{v}$ by the twinning shear, then

$$(M\ P\ M)[M;\underline{u}] = [M;\underline{v}] \quad\quad\quad .....(15b)$$

and if $\underline{h}$ is a plane normal which after deformation becomes $\underline{k}$, then

$$(\underline{h};M^*)(M\ P\ M)^{-1} = (\underline{k};M^*) \quad\quad\quad .....(15c)$$

These laws can be used to verify that $\underline{p}$ and $\underline{d}$ are unaffected by the twinning shear, and that the magnitude of a vector originally along η_2 is not changed by the deformation; similarly, the spacing of the planes initially parallel to K_2 remains the same after deformation, although the planes are rotated.

The Concept of a Correspondence Matrix

The property of the homogeneous deformations we have been considering is that points which are initially colinear remain so in spite of the deformation, and lines which are initially coplanar remain coplanar after the strain. Using the data of example 13, it can easily be verified that the deformation (M P M) alters the vector $[M;\underline{u}] = [0\ 0\ 1]$ to a new vector $[M;\underline{v}] = (1/6)[1\ 1\ 4]$

i.e., $$(M \ P \ M)[0 \ 0 \ 1]_M = (1/6)[1 \ 1 \ 4]_M \qquad .....(15d)$$

The indices of this new vector $\underline{v}$ relative to the twin basis T can be obtained using the co-ordinate transformation matrix (T J M), so that

$$(T \ J \ M)(1/6)[1 \ 1 \ 4]_M = [T;\underline{v}] = (1/2)[\bar{1} \ \bar{1} \ 0]_T \qquad .....(15e)$$

Hence, the effect of the shear stress is to deform a vector $[0 \ 0 \ 1]_M$ of the parent lattice into a vector $(1/2)[\bar{1} \ \bar{1} \ 0]_T$ of the twin. Equations 15d and 15e could have been combined to obtain this result, as follows:

$$(T \ J \ M)(M \ P \ M)[M;\underline{u}] = [T;\underline{v}] \qquad .....(15f)$$

or

$$(T \ C \ M)[M;\underline{u}] = [T;\underline{v}] \qquad .....(15g)$$

where

$$(T \ J \ M)(M \ P \ M) = (T \ C \ M) \qquad .....(15h)$$

The matrix (T C M) is called the correspondence matrix; the initial vector $\underline{u}$ in the parent basis, due to deformation becomes a *corresponding* vector $\underline{v}$ with indices $[T;\underline{v}]$ in the twin basis. The correspondence matrix tells us that a certain vector in the twin is formed by deforming a particular corresponding vector of the parent. In the example considered above, the vector $\underline{u}$ has rational components in M (i.e., the components are small integers or fractions) and $\underline{v}$ has rational components in T. It follows that the elements of the correspondence matrix (T C M) must also be rational numbers or fractions. The correspondence matrix can usually be written from inspection since its columns are rational lattice vectors referred to the second basis produced by the deformation from the basis vectors of the first basis.

We can similarly relate planes in the twin to planes in the parent, the correspondence matrix being given by

$$(M \ C \ T) = (M \ P \ M)^{-1}(M \ J \ T) \qquad .....(15i)$$

where

$$(\underline{h};M^*)(M \ C \ T) = (\underline{k};T^*)$$

so that the plane $(\underline{k};T^*)$ of the twin was formed by the deformation of the plane $(\underline{h};M^*)$ of the parent.

Stepped Interfaces

A planar coherent twin boundary (unit normal $\underline{p}$) can be generated from a single crystal by shearing on the twinning plane $\underline{p}$, the unit shear direction and shear magnitude being $\underline{d}$ and m respectively.

On the other hand, to generate a similar boundary but containing a step of height h requires [24,27,31] additional virtual operations (Fig. 14). The single crystal is first slit along a plane which is not parallel to $\underline{p}$ (Fig. 14b), before applying the twinning shear. The shear which generates the twinned orientation also opens up the slit (Fig. 14c), which then has to be rewelded (Fig. 14d) along the original cut; this produces the required stepped interface. A Burgers circuit constructed around the stepped interface will, when compared with an equivalent circuit constructed around the unstepped interface exhibit a closure failure. This closure failure gives the Burgers vector $\underline{b}_I$ associated with the step:

$$\underline{b}_I = hm\underline{d} \qquad ....(16a)$$

The operations outlined above indicate one way of generating the required stepped interface. They are simply the virtual operations which allow us to produce the required defect - similar operations were first used by Volterra [32] in describing the elastic properties of cut and deformed cylinders, operations which were later recognised to generate the ordinary dislocations that metallurgists are so familiar with.

Having defined $\underline{b}_I$, we note that an initially planar coherent twin boundary can acquire a step if a dislocation of Burgers vector $\underline{b}_m$ crosses the interface. The height of the step is given by [31]

34

$$h = \underline{b}_m \cdot \underline{p}$$

so that $\qquad\qquad \underline{b}_I = m(\underline{b}_m \cdot \underline{p})\underline{d}$ $\qquad\qquad$ .....(16b)

From eq.11d, the invariant plane strain necessary to generate the twin from the parent lattice is given by $(M\ P\ M) = I + m[M;\underline{d}](\underline{p};M^*)$ so that eq.16b becomes

$$[M;\underline{b}_I] = [\underline{b}_m;M] - (M\ P\ M)[M;\underline{b}_m] \qquad\qquad .....(16c)$$

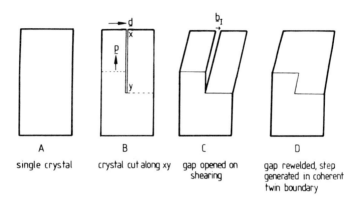

A	B	C	D
single crystal	crystal cut along xy	gap opened on shearing	gap rewelded, step generated in coherent twin boundary

<u>Fig. 14</u>   The virtual operations (Ref.27) used in determining $\underline{b}_I$.

<u>Example 14: Interaction of Dislocations with Interfaces</u>

Deduce the correspondence matrix for the deformation twin discussed in example 13 and hence show that there are no geometrical restrictions to the passage of slip dislocations across coherent twin boundaries in FCC materials.

From example 13,

$$(T\ J\ M) = (1/3)\begin{pmatrix} 1 & \bar{2} & \bar{2} \\ \bar{2} & 1 & \bar{2} \\ \bar{2} & \bar{2} & 1 \end{pmatrix} \qquad \text{and} \qquad (M\ P\ M) = (1/6)\begin{pmatrix} 7 & 1 & 1 \\ 1 & 7 & 1 \\ \bar{2} & \bar{2} & 4 \end{pmatrix}$$

The correspondence matrix $(T\ C\ M)$ which associates each vector of the parent with a corresponding vector in the twin is, from eq.15h, given by

$$(T\ C\ M) = (T\ J\ M)(M\ P\ M)$$
$$(M\ C\ T) = (M\ P\ M)^{-1}(M\ J\ T)$$

so that

$$(T\ C\ M) = (M\ C\ T) = (1/2)\begin{pmatrix} 1 & \bar{1} & \bar{1} \\ \bar{1} & 1 & \bar{1} \\ \bar{2} & \bar{2} & 0 \end{pmatrix}$$

The character of a dislocation will in general be altered on crossing an interface. This is because the crossing process introduces a step in the interface, rather like the slip steps which arise at the free surfaces of deformed crystals. We consider the case where a dislocation crosses a coherent twin boundary. The interfacial step has dislocation character so that the original dislocation (Burgers vector $\underline{b}_m$) from the parent crystal is in effect converted into *two* dislocations, one being the step (Burgers vector $\underline{b}_I$) and the other the dislocation (Burgers vector $\underline{b}_t$) which has penetrated the interface and entered into the twin lattice. If the total Burgers vector content of the system is to be preserved then it follows that in general, $\underline{b}_t \neq \underline{b}_m$, since $\underline{b}_m = \underline{b}_I + \underline{b}_t$. Using this equation and eq.16c, we see that

$$[M;\underline{b}_t] = (M\ P\ M)[M;\underline{b}_m]$$

or

$$[T;\underline{b}_t] = (T\ J\ M)(M\ P\ M)[M;\underline{b}_m]$$

so that

$$[T;\underline{b}_t] = (T\ C\ M)[M;\underline{b}_m] \qquad .....(17a)$$

Clearly, dislocation glide across the coherent interface will not be hindered if $\underline{b}_t$ is a perfect lattice vector of the twin. If this is not the case and $\underline{b}_t$ is a partial dislocation in the twin, then glide across the interface will be hindered because the motion of $\underline{b}_t$ in the twin would leave a stacking fault trailing all the way from the interface to the position of the partial dislocation in the twin.

There is an additional condition to be fulfilled for easy glide across the interface; the *corresponding* glide planes $\underline{p}_m$ and $\underline{p}_t$ of dislocations $\underline{b}_m$ and $\underline{b}_t$ in the parent and twin lattices respectively, must meet edge to edge in the interface. Now,

$$(\underline{p}_t; T^*) = (\underline{p}_m; M^*)(M\ C\ T) \qquad .....(17b)$$

If the interface plane normal is $\underline{p}_I$, then the edge to edge condition is satisfied if $\underline{p}_m \wedge \underline{p}_I \| \underline{p}_t \wedge \underline{p}_I$.

Dislocations in FCC materials usually glide on close-packed {1 1 1} planes and have Burgers vectors of type $(a/2)\langle 1\ \bar{1}\ 0\rangle$. Using the data of Table 1 it can easily be verified that all the close-packed planes of the parent lattice meet the corresponding glide planes in the twin edge to edge in the interface, which is taken to be the coherent $(1\ 1\ 1)_M$ twinning plane. Furthermore, all the $(a/2)\langle 1\ \bar{1}\ 0\rangle$ Burgers vectors of glide dislocations in the parent correspond to perfect lattice dislocations in the twin. It must be concluded that the coherent twin boundary for {1 1 1} twins in FCC metals does not offer any geometrical restrictions to the transfer of slip between the parent and product lattices.

Table 1: Corresponding Glide Planes and Burgers Vectors

Parent	Twin
$(a/2)[1\ 1\ 0]$	$(\ a\ )[0\ 0\ \bar{1}]$
$(a/2)[1\ 0\ 1]$	$(a/2)[0\ \bar{1}\ \bar{1}]$
$(a/2)[0\ 1\ 1]$	$(a/2)[\bar{1}\ 0\ \bar{1}]$
$(a/2)[1\ \bar{1}\ 0]$	$(a/2)[1\ \bar{1}\ 0]$
$(a/2)[1\ 0\ \bar{1}]$	$(a/2)[1\ 0\ \bar{1}]$
$(a/2)[0\ 1\ \bar{1}]$	$(a/2)[0\ 1\ \bar{1}]$
$(1\ 1\ 1)$	$(\bar{1}\ \bar{1}\ \bar{1})$
$(1\ 1\ \bar{1})$	$(1\ 1\ \bar{1})$
$(1\ \bar{1}\ 1)$	$(0\ \bar{2}\ 0)$
$(\bar{1}\ 1\ 1)$	$(\bar{2}\ 0\ 0)$

These data (Table 1) also show that all dislocations with Burgers vectors in the $(1\ 1\ 1)_M$ plane are unaffected, both in magnitude and direction, as a result of crossing into the twin. For example, $(a/2)[1\ \bar{1}\ 0]_M$ becomes $(a/2)[1\ \bar{1}\ 0]_T$ so that $|\underline{b}_m| = |\underline{b}_t|$, and using $(T\ J\ M)$ it can be demonstrated that $[1\ \bar{1}\ 0]_T \| [1\ \bar{1}\ 0]_T$. This result is expected because these particular dislocations cannot generate a step in the $(1\ 1\ 1)_M$ interface when they cross into the twin lattice (see eq.16b). Only dislocations with Burgers vectors not parallel to the interface cause the formation of steps.

The data further illustrate the fact that when $\underline{b}_m$ lies in the $(1\ 1\ 1)_M$ plane, there is no increase in energy due to the reaction $\underline{b}_m \rightarrow \underline{b}_I + \underline{b}_t$, which occurs when a dislocation crosses the interface. This is because $\underline{b}_I = 0$ and $\underline{b}_t = \underline{b}_m$. For all other cases $\underline{b}_I$ is not zero and since $|\underline{b}_I|$ is never less than $|\underline{b}_m|$, $\underline{b}_m \rightarrow \underline{b}_I + \underline{b}_t$ is always energetically unfavourable. In fact, in the example being discussed, there can never be an energy reduction when an $(a/2)<1\ \overline{1}\ 0>$ dislocation penetrates the coherent twin boundary. The dislocations cannot therefore spontaneously cross the boundary. A trivial case where dislocations might spontaneously cross a boundary is when the latter is a free surface, assuming that the increase in surface area (and hence surface energy) due to the formation of a step is not prohibitive. Spontaneous penetration of the interface might also become favourable if the interface separates crystals with very different elastic properties.

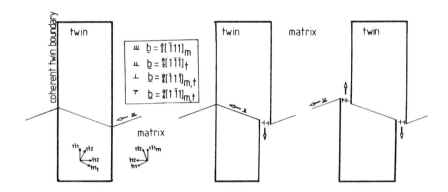

Fig. 15   The passage of a slip dislocation across a coherent twin boundary in a BCC crystal. The twinning system is $\{\overline{1}\ 1\ 2\}<1\ \overline{1}\ 1>$, $s = 2^{-1/2}$. The subscripts m and t refer to the twin and matrix respectively; the open arrows indicate the sense of the Burgers vectors and the dislocation line vectors are all parallel to $[\overline{1}\ \overline{1}\ 0]_{M,T}$.

The results obtained show that single dislocations can glide into twins in FCC crystals without leaving a fault; there are no geometrical restrictions to the passage of slip dislocations across the coherent twin boundaries concerned. It can similarly be demonstrated that slip dislocations can comfortably traverse the coherent twin boundaries of $\{1\ 1\ 2\}$ twins in BCC or BCT lattices and this has implications on the interpretation of the strength of martensite [31]. As will be discussed later, the substructure of martensite plates in steels (and in many non-ferrous alloys) often consists of very finely spaced $\{1\ 1\ 2\}$ transformation twins. It was at one time believed that the twins were mainly responsible for the high strength of ferrous martensites, because the numerous twin boundaries should hinder slip - the analysis above clearly suggests otherwise. Indeed, twinned martensites which do not contain carbon also do not exhibit exceptionally high strengths and it is now generally accepted that the strength of virgin ferrous martensites is largely due to interstitial solid solution hardening by carbon atoms, or in the case of lightly autotempered martensites due to carbon atom clustering or fine precipitation. Consistent with this, it is found that Fe-30Ni (wt.pct.) twinned martensites are not particularly hard.

Finally, it should be mentioned that even when glide across coherent twin boundaries in martensites should be unhindered, the boundaries will cause a small amount of hardening, partly because the corresponding slip systems in the matrix and twin will in general be differently stressed [31,33] (simply because they are not necessarily parallel) and partly due to the work necessary to create the steps in the interfaces. It is emphasized, however, that these should be relatively small contributions to the strength of martensite. Fig. 15 illustrates the passage of a slip dislocation across a coherent {1 1 2} twin interface in a BCC material.

Eigenvectors and Eigenvalues

In Chapter 2, eq.8b was used to determine the direction which remains unrotated and undistorted as a result of a rigid body rotation. To examine the properties of invariant-plane strains and other strains (or linear transformations) in more detail, it is necessary to establish a more general method of determining the directions which remain *unrotated*, though not necessarily undistorted, as a consequence of the deformation concerned. Vectors lying along such unrotated directions are called *eigenvectors* of the deformation (or transformation) matrix, and the ratios of their final to initial lengths are the corresponding *eigenvalues* of the matrix. Considering the deformation matrix (A S A), the unrotated directions may be determined by solving the equations

$$(A\ S\ A)[A;\underline{u}] = \lambda[A;\underline{u}] \qquad .....(18a)$$

where $\underline{u}$ is a unit vector lying along an eigenvector, A is a convenient orthonormal basis and λ is a scalar quantity. This equation shows that the vector $\underline{u}$ does not change in direction as a result of (A S A), although its length changes by the ratio λ (eq.18a can be compared with eq.8b, where $\lambda = 1$). If I is a 3x3 identity matrix, then on rearranging eq.18a, we obtain

$$\{(A\ S\ A) - \lambda I\}[A;\underline{u}] = 0 \qquad .....(18b)$$

which can be written more fully as:

$$\begin{pmatrix} S_{11}-\lambda & S_{12} & S_{13} \\ S_{21} & S_{22}-\lambda & S_{23} \\ S_{31} & S_{32} & S_{33}-\lambda \end{pmatrix} \begin{pmatrix} u_1 \\ u_2 \\ u_3 \end{pmatrix} = 0 \qquad .....(18c)$$

where $[A;\underline{u}] = [u_1\ u_2\ u_3]$. This system of homogeneous equations has non-trivial solutions if

$$\begin{vmatrix} S_{11}-\lambda & S_{12} & S_{13} \\ S_{21} & S_{22}-\lambda & S_{23} \\ S_{31} & S_{32} & S_{33}-\lambda \end{vmatrix} = 0 \qquad .....(18d)$$

The expansion of this determinant yields an equation which is in general cubic in λ; the roots of this equation are the three eigenvalues λ_i. Associated with each of the eigenvalues is a corresponding eigenvector whose components may be obtained by substituting each eigenvalue, in turn, into eq.18c. Of course, since every vector which lies along the unrotated direction is an eigenvector, if $\underline{u}$ is a solution of eq.18c then $k\underline{u}$ must also satisfy eq.18c, k being a scalar constant. If the matrix (A S A) is real then there must exist three eigenvalues, at least one of which is necessarily real. If (A S A)

38

SLIP, TWINNING AND OTHER INVARIANT-PLANE STRAINS

is symmetrical then all three of its eigenvalues are real; the existence of three real eigenvalues does not however imply that the deformation matrix is symmetrical. Every real eigenvalue implies the existence of a corresponding vector which remains unchanged in direction as a result of the operation of (A S A).

Example 15: Eigenvectors and Eigenvalues

Find the eigenvalues and eigenvectors of

$$(A \ S \ A) = \begin{pmatrix} 18 & -6 & -6 \\ -6 & 21 & 3 \\ -6 & 3 & 21 \end{pmatrix}$$

To solve for the eigenvalues, we use eq.18d to form the determinant

$$\begin{vmatrix} 18-\lambda & -6 & -6 \\ -6 & 21-\lambda & 3 \\ -6 & 3 & 21-\lambda \end{vmatrix} = 0$$

which on expansion gives the cubic equation

$$(12 - \lambda)(\lambda - 30)(\lambda - 18) = 0$$

with the roots

$$\lambda_1 = 12, \lambda_2 = 30 \text{ and } \lambda_3 = 18$$

To find the eigenvector $\underline{u} = [A;\underline{u}] = [u_1 \ u_2 \ u_3]$ corresponding to λ_1, we substitute λ_1 into eq.18c to obtain

$$\begin{aligned} 6u_1 - 6u_2 - 6u_3 &= 0 \\ -6u_1 + 9u_2 + 3u_3 &= 0 \\ -6u_1 + 3u_2 + 9u_3 &= 0 \end{aligned}$$

These equations can be simultaneously solved to show that $u_1 = 2u_2 = 2u_3$. The other two eigenvectors, $\underline{v}$ and $\underline{w}$, corresponding to λ_2 and λ_3 respectively, can be determined in a similar way. Hence, it is found that:

$$[A;\underline{u}] = (6^{-1/2})[2 \ 1 \ 1]$$
$$[A;\underline{v}] = (3^{-1/2})[\bar{1} \ 1 \ 1]$$
$$[A;\underline{w}] = (2^{-1/2})[0 \ 1 \ \bar{1}]$$

All vectors parallel to $\underline{u}$, $\underline{v}$ or $\underline{w}$ remain unchanged in direction, though not in magnitude, due to the deformation (A S A).

Comments
(i) Since the matrix (A S A) is symmetrical, we find three real eigenvectors, which form an orthogonal set.
(ii) A negative eigenvalue implies that a vector initially parallel to the corresponding eigenvector becomes antiparallel (changes sign) on deformation. A deformation like this is physically impossible.
(iii) If a new orthonormal basis B is defined, consisting of unit basis vectors parallel to $\underline{u}$, $\underline{v}$ and $\underline{w}$ respectively, then the deformation (A S A) can be expressed in the new basis with the help of a similarity transformation. From eq.11,

$$(B \ S \ B) = (B \ J \ A)(A \ S \ A)(A \ J \ B) \qquad \qquad .....(18e)$$

where the columns of (A J B) consist of the components (referred to the basis A) of the eigenvectors $\underline{u}$, $\underline{v}$ and $\underline{w}$ respectively, so that

39

$$(\text{B S B})= \begin{pmatrix} u_1 & u_2 & u_3 \\ v_1 & v_2 & v_3 \\ w_1 & w_2 & w_3 \end{pmatrix} \begin{pmatrix} 18 & -6 & -6 \\ -6 & 3 & 21 \\ -6 & 21 & 3 \end{pmatrix} \begin{pmatrix} u_1 & v_1 & w_1 \\ u_2 & v_2 & w_2 \\ u_3 & v_3 & w_3 \end{pmatrix} = \begin{pmatrix} 18 & 0 & 0 \\ 0 & 30 & 0 \\ 0 & 0 & 12 \end{pmatrix}$$

Notice that (B S B) is a diagonal matrix (off diagonal terms equal zero) because it is referred to a basis formed by the principal axes of the deformation - i.e., the three orthogonal eigenvectors. The matrix representing the Bain Strain in chapter 1 is also diagonal because it is referred to the principal axes of the strain. Any real symmetrical matrix can be diagonalised using the procedure illustrated above.

(B S B) is called the 'diagonal' representation of the deformation (since off diagonal components are zero) and this special representation will henceforth be identified by placing a bar over the matrix symbol: (B $\bar{\text{S}}$ B).

Stretch and Rotation

Inspection of the invariant-plane strain (Z P1 Z) illustrated in Fig. 10a shows that it is possible to find three initially orthogonal axes which are not rotated by the deformation. These *principal* axes are the eigenvectors of (Z P1 Z); any two mutually perpendicular axes in the invariant-plane constitute two of the eigenvectors and the third is parallel to the invariant-plane normal. The matrix (Z P1 Z) is symmetrical (eq.11d) and indeed, would have to be symmetrical to yield three real and orthogonal eigenvectors. Since all vectors lying in the invariant-plane are unaffected by the deformation, two of the eigenvectors have eigenvalues of unity; the third has the eigenvalue $(1+\delta)$. Hence the deformation simply consists of an extension along one of the principal axes.

As discussed in Chapter 1 (under Homogeneous deformations), a strain like (Z P1 Z) is called a *pure deformation* and has the following characteristics:
(i) It has a symmetrical matrix representation irrespective of the choice of basis.
(ii) It consists of simple extensions or contractions along the principal axes. The ratios of the final to initial lengths of vectors parallel to the principal axes are called the principal deformations, and the change in length per unit length the principal strains.
(iii) It is possible to find three real and orthogonal eigenvectors.

We note that a pure deformation need not be an invariant-plane strain; the strain (A S A) of example 15 is a pure deformation, as is the Bain strain.

On the other hand, the shear (Z P2 Z) illustrated in Fig. 10b is not a pure deformation because it is only possible to identify two mutually perpendicular eigenvectors, both of which must lie in the invariant-plane. All other vectors are rotated by the shearing action. The deformation is illustrated again in Fig. 16a, where the original lattice, represented as a sphere, is sheared into an ellipsoid. The invariant-plane of the deformation contains the z1 and z2 axes. The deformation can be *imagined* to occur in two stages, the first one involving simple extensions and contractions along the y1 and y3 directions respectively (Fig. 16b) and the second involving a rigid body rotation of the ellipsoid, about the axis z2lly2, through a right-handed angle ϕ.

In essence, we have just carried out an imaginary factorisation of the impure strain (Z P2 Z) into a pure strain (Fig. 16b) and a rigid body rotation. If the pure strain part is referred to as (Z Q Z) and the rotation part as (Z J Z), then

$$(\text{Z P2 Z}) = (\text{Z J Z})(\text{Z Q Z}) \qquad \qquad .....(19a)$$

It was arbitrarily chosen that the pure strain would occur first and be followed by the rigid body rotation, but the reverse order is equally acceptable,

40

$$(Z\ P2\ Z) = (Z\ Q2\ Z)(Z\ J2\ Z)$$

where in general, $(Z\ Q2\ Z) \neq (Z\ Q\ Z)$ and $(Z\ J2\ Z) \neq (Z\ J\ Z)$

In general, any real deformation can be factorised into a pure strain and a rigid body rotation, but it is important to realise that the factorisation is simply a mathematical convenience and the deformation does not actually occur in the two stages. The factorisation in no way indicates the path by which the initial state reaches the final state and is merely phenomenological.

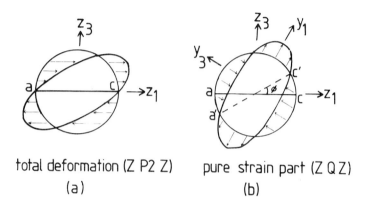

total deformation $(Z\ P2\ Z)$ pure strain part $(Z\ Q\ Z)$

(a)   (b)

Fig. 16   Factorisation of a simple shear $(Z\ P2\ Z)$ into a pure deformation $(Z\ Q\ Z)$ and a right handed rigid body rotation of ϕ about $z3$. In (a), ac is the trace of the invariant plane. $(Z\ Q\ Z)$ leaves ac undistorted but rotated to a'c' and rigid body rotation brings a'c' into coincidence with ac. The axes y_1, y_2 and y_3 are the principal axes of the pure deformation. The undeformed shape is represented as a sphere in three dimensions.

The actual factorisation can be considered in terms of the arbitrary deformation $(Z\ S\ Z)$, referred to an orthonormal basis Z. Bearing in mind that $(Z\ S'\ Z)$ is the transpose of $(Z\ S\ Z)$,

$$(Z\ S'\ Z)(Z\ S\ Z) = (Z\ Q'\ Z)(Z\ J'\ Z)(Z\ J\ Z)(Z\ Q\ Z)$$

or $$(Z\ S'\ Z)(Z\ S\ Z) = (Z\ Q\ Z)^2 \qquad \qquad .....(19b)$$

since $(Z\ J'\ Z)(Z\ J\ Z) = I$ and $(Z\ Q'\ Z) = (Z\ Q\ Z)$, $(Z\ Q\ Z)$ being a pure deformation having a symmetrical matrix representation. If the product $(Z\ S'\ Z)(Z\ S\ Z)$ is written as the symmetrical matrix $(Z\ T\ Z)$, then the eigenvalues λ_i of $(Z\ T\ Z)$ are also the eigenvalues of $\{(Z\ Q\ Z)^2\}$, so that the eigenvalues of $(Z\ Q\ Z)$ are $\lambda_i^{1/2}$. If the eigenvectors of $(Z\ T\ Z)$ are $\underline{u}, \underline{v}$ and $\underline{w}$ (corresponding to λ_1, λ_2 and λ_3 respectively), then $(Z\ T\ Z)$ can be diagonalised by similarity transforming it to another orthonormal basis Y formed by the vectors $\underline{u}, \underline{v}$ and $\underline{w}$. From eq.18e,

$$(Y\ \bar{T}\ Y) = \begin{pmatrix} \lambda_1 & 0 & 0 \\ 0 & \lambda_2 & 0 \\ 0 & 0 & \lambda_3 \end{pmatrix} = (Y\ J\ Z)(Z\ T\ Z)(Z\ J\ Y)$$

where the columns of $(Z\ J\ Y)$ consist of the components of $\underline{u}, \underline{v}$ and $\underline{w}$, respectively, when the latter

41

are referred to the Z basis. Since $(Y \bar{Q} Y)^2 = (Y \bar{T} Y)$, $(Z\ Q\ Z) = (Z\ J\ Y)(Y\ \bar{T}\ Y)^{1/2}(Y\ J\ Z)$ where the square root of a diagonal matrix $(Y\ \bar{T}\ Y)$ is such that $(Y\ \bar{T}\ Y)^{1/2}(Y\ \bar{T}\ Y)^{1/2} = (Y\ \bar{T}\ Y)$. It follows that:

$$(Z\ Q\ Z) = \begin{pmatrix} u_1 & v_1 & w_1 \\ u_2 & v_2 & w_2 \\ u_3 & v_3 & w_3 \end{pmatrix} \begin{pmatrix} \lambda_1^{1/2} & 0 & 0 \\ 0 & \lambda_2^{1/2} & 0 \\ 0 & 0 & \lambda_3^{1/2} \end{pmatrix} \begin{pmatrix} u_1 & u_2 & u_3 \\ v_1 & v_2 & v_3 \\ w_1 & w_2 & w_3 \end{pmatrix}$$

.....(19c)

It is worth repeating that in eq.19c, λ_i are the eigenvectors of the matrix $\{(Z\ S'\ Z)(Z\ S\ Z)\}$ and u_i, v_i and w_i are the components, in the basis Z of the eigenvectors of $\{(Z\ S'\ Z)(Z\ S\ Z)\}$. The rotation part of the strain $(Z\ S\ Z)$ is simply

$$(Z\ J\ Z) = (Z\ S\ Z)(Z\ Q\ Z)^{-1}$$

.....(19d)

Example 16: The FCC to HCP transformation revisited

A Co-6.5wt.pct.Fe alloy transforms from an FCC (γ) structure to a HCP martensite structure with virtually zero change in density [21]. The invariant plane of the transformation is the close-packed $\{1\ 1\ 1\}_\gamma$ plane, the shear direction being $<1\ 1\ \bar{2}>_\gamma$. The magnitude of the shear is $8^{-1/2}$, which is half the normal twinning shear for FCC crystals. By factorising the total transformation strain into a pure strain and a rigid body rotation, show that the maximum extension or contraction suffered by any vector of the parent lattice, as a result of the transformation, is less than 20%.

Representing the FCC parent lattice in an orthonormal basis Z, consisting of unit basis vectors parallel to [1 0 0], [0 1 0] and [0 0 1] FCC directions respectively, and substituting $(p;Z^*) = (3^{-1/2})(1\ 1\ 1)$, $[Z;\underline{d}] = (6^{-1/2})[1\ 1\ \bar{2}]$ and $m = 8^{-1/2}$ into eq.11d, the total transformation strain $(Z\ P\ Z)$ is found to be:

$$(Z\ P\ Z) = (1/12) \begin{pmatrix} 13 & 1 & -2 \\ 1 & 13 & 1 \\ -2 & -2 & 10 \end{pmatrix}$$

This can be factorised into a pure strain $(Z\ Q\ Z)$ and a rigid body rotation $(Z\ J\ Z)$. That the eigenvectors of $(Z\ Q\ Z)$ represent the directions along which the maximum length changes occur can be seen from Fig. 6 (the basis Y of Fig. 6 differs from the present basis Z. In fact, the basis vectors of Y are parallel to the eigenvectors of $(Z\ Q\ Z)$). $(Z\ P\ Z)$ is illustrated in Fig. 6c and $(Z\ Q\ Z)$ in Figs. 6a,b. It is seen that the axes of the ellipsoid represent directions along which the greatest length changes occur; these axes are of course the eigenvectors of $(Z\ Q\ Z)$. Writing $(Z\ T\ Z) = (Z\ P'\ Z)(Z\ P\ Z)$, we obtain:

$$(Z\ T\ Z) = (1/144) \begin{pmatrix} 174 & 30 & -6 \\ 30 & 174 & -6 \\ -6 & -6 & 102 \end{pmatrix}$$

The eigenvalues and eigenvectors of $(Z\ T\ Z)$ are:

$\lambda_1 = 1.421535$ $[Z;\underline{u}] = [\ 0.704706 \quad 0.704706 \quad -0.082341]$

42

$\lambda_2 = 1.000000$ $[Z{:}\underline{v}] = [\ 0.707107 \quad -0.707107 \quad 0.000000]$

$\lambda_3 = 0.703465$ $[Z{;}\underline{w}] = [\ 0.058224 \quad 0.058224 \quad 0.996604]$

Notice that the eigenvectors form an orthogonal set and that consistent with the fact that $\underline{v}$ lies in the invariant plane, λ_2 has a value of unity. $\underline{u}$, $\underline{v}$ and $\underline{w}$ are also the eigenvectors of $(Z\ Q\ Z)$. The eigenvalues of $(Z\ Q\ Z)$ are given by the square roots of the eigenvalues of $(Z\ T\ Z)$; they are 1.192282, 1.0 and 0.838728. Hence, the maximum extensions and contractions are less than 20% since each eigenvalue is the ratio of the final to initial length of a vector parallel to an eigenvector. The maximum extension occurs along $\underline{u}$ and the maximum contraction along $\underline{w}$. The matrix $(Z\ Q\ Z)$ is given by eq.19c as:

$(Z\ Q\ Z) =$

$$\begin{pmatrix} 0.70471 & 0.70711 & 0.05822 \\ 0.70471 & -0.70711 & 0.05822 \\ -0.08234 & 0.00000 & 0.99660 \end{pmatrix} \begin{pmatrix} 1.19228 & 0.0 & 0.0 \\ 0.0 & 1.00000 & 0.0 \\ 0.0 & 0.0 & 0.83873 \end{pmatrix} \begin{pmatrix} 0.70471 & 0.70471 & -0.08234 \\ 0.70711 & -0.70711 & 0.00000 \\ 0.05822 & 0.05822 & 0.99660 \end{pmatrix}$$

$$(Z\ Q\ Z) = \begin{pmatrix} 1.094944 & 0.094943 & -0.020515 \\ 0.094943 & 1.094944 & -0.020515 \\ -0.020515 & -0.020515 & 0.841125 \end{pmatrix}$$

and

$$(Z\ Q\ Z)^{-1} = \begin{pmatrix} 0.920562 & -0.079438 & 0.020515 \\ -0.079438 & 0.920562 & 0.020515 \\ 0.020515 & 0.020515 & 1.189885 \end{pmatrix}$$

From eq.19d, $(Z\ J\ Z) = (Z\ P\ Z)(Z\ Q\ Z)^{-1}$

$$(Z\ J\ Z) = \begin{pmatrix} 0.992365 & -0.007635 & 0.123091 \\ -0.007635 & 0.992365 & 0.123091 \\ -0.123092 & -0.123092 & 0.984732 \end{pmatrix}$$

The matrix $(Z\ J\ Z)$, from eq.8, represents a right-handed rotation of 10.03^{o} about $[1\ \bar{1}\ 0]_Z$ axis.

It is interesting to examine what happens to the vector $[1\ 1\ \bar{2}]_Z$ due to the operations $(Z\ Q\ Z)$ and $(Z\ J\ Z)$:

$$(Z\ Q\ Z)[1\ 1\ \bar{2}]_Z = [1.230916\ 1.230916\ -1.723280]_Z$$

where the new vector can be shown to have the same magnitude as $[1\ 1\ \bar{2}]$ but points in a different direction. The effect of the pure rotation is

$$(Z\ J\ Z)[1.230916\ 1.230916\ -1.723280]_Z = [1\ 1\ \bar{2}]_Z$$

Thus, the pure strain deforms $[1\ 1\ \bar{2}]_Z$ into another vector of identical magnitude and the pure rotation brings this new vector back into the $[1\ 1\ \bar{2}]_Z$ direction, the net operation leaving it invariant, as expected, since $[1\ 1\ \bar{2}]_Z$ is the shear direction which lies in the invariant plane. Referring to Fig. 6, the direction fe = $[1\ \bar{1}\ 0]_Z$, cd = $[1\ 1\ \bar{2}]_Z$ and c'd' = $[1.230916\ 1.230916\ -1.723280]_Z$. c'd' is brought into coincidence with cd by the rigid body rotation $(Z\ J\ Z)$ to generate the invariant plane containing fe and cd.

43

Physically, the FCC to HCP transformation occurs by the movement of a single set of Shockley partial dislocations, Burgers vector $\underline{b}$ = $(a/6)<1\ 1\ \overline{2}>_\gamma$ on *alternate* close-packed $\{1\ 1\ 1\}_\gamma$ planes. To produce a fair thickness of HCP martensite, a mechanism has to be sought which allows Shockley partials to be generated on every other slip plane. Some kind of a pole mechanism (see for example, p.310 of ref.25) would allow this to happen, but there is as yet no experimental evidence confirming this. Motion of the partials would cause a shearing of the γ lattice, on the system $\{1\ 1\ 1\}_\gamma$ $<1\ 1\ \overline{2}>_\gamma$, the average magnitude $\overline{s}$ of the shear being $\overline{s}$ = $|\underline{b}|/2d$, where d is the spacing of the close-packed planes. Hence, $\overline{s}$ = $(6^{-1/2}a)/(2(3^{-1/2}a))$ = $8^{-1/2}$. This is exactly the shear system we used in generating the matrix (Z P Z) and the physical effect of the shear on the shape of an originally flat surface is, in general, to tilt the surface (about a line given by its intersection with the HCP habit plane) through some angle dependent on the indices of the free surface. By measuring such tilts it is possible to deduce $\overline{s}$, which has been experimentally confirmed to equal half the twinning shear.

In FCC crystals, the close-packed planes have a stacking sequence ...ABCABCABC...; the passage of a single Shockley partial causes the sequence to change to ...ABA... creating a three layer thick region of HCP phase since the stacking sequence of close-packed planes in the HCP lattice has a periodicity of 2. This then is the physical manner in which the transformation occurs, the martensite having a $\{1\ 1\ 1\}_\gamma$ habit plane - if the parent product interface deviates slightly from $\{1\ 1\ 1\}_\gamma$ then it will consist of stepped sections of close-packed plane, the steps representing the Shockley partial transformation dislocations. The spacing of the partials along $<1\ 1\ 1>_\gamma$ would be 2d. In other words, in the stacking sequence ABC, the motion of a partial on B would leave A and B unaffected though C would be displaced by $(a6^{-1/2})<1\ 1\ \overline{2}>_\gamma$ to a new position A, giving ABA stacking. Partials could thus be located on every alternate plane of the FCC crystal.

Hence, we see that the matrix (Z P Z) is quite compatible with the microscopic dislocation based mechanism of transformation. (Z P Z) predicts the correct macroscopic surface relief effect and its invariant plane is the habit plane of the martensite. However, if (Z P Z) is considered to act homogeneously over the entire crystal, then it would carry half the atoms into the wrong positions. For instance, if the habit plane is designated A in the sequence ABC of close packed planes, then the effect of (Z P Z) is to leave A unchanged, shift the atoms on plane C by $2\overline{s}d$ and those on plane B by $\overline{s}d$ along $<1\ 1\ \overline{2}>_\gamma$. Of course, this puts the atoms originally in C sites into A sites, as required for HCP stacking. However, the B atoms are located at positions half way between B and C sites, through a distance $(a/12)<1\ 1\ \overline{2}>_\gamma$. *Shuffles* are thus necessary to bring these atoms back into the original B positions and to restore the ...ABA... HCP sequence. These atomic movements in the middle layer are called shuffles because they occur through very small distances (always less than the interatomic spacing) and do not affect the macroscopic shape change [27]. The shuffle here is a purely formal concept; consistent with the fact that the Shockley partials glide over alternate close-packed planes, the deformation (Z P Z) must in fact be considered homogeneous only on a scale of every two planes. By locking the close-packed planes together in pairs, we avoid displacing the B atoms to the wrong positions and thus automatically avoid the reverse shuffle displacement.

In the particular example discussed above, the dislocation mechanism is established experimentally and physically reasonable shear systems were used in determining (Z P Z). However, in general it is possible to find an infinite number of deformations [5,27] which may accomplish the same lattice change and slightly empirical criteria have to be used in selecting the correct deformation. One such criterion could involve the selection of deformations which involve the minimum principal strains and the minimum degree of shuffling, but intuition and experimental evidence is almost always necessary to reach a decision.

The Bain strain which transforms the FCC lattice to the BCC lattice is believed to be the correct choice because it seems to involve the least atomic displacements and zero shuffling of atoms [34]. The absence of shuffles can be deduced from the Bain correspondence matrix (α C γ) which can be deduced from inspection since its columns are rational lattice vectors referred to the α basis, produced by the deformation of the basis vectors of the γ basis; since [1 0 0]$_\gamma$ is deformed to [1 1 0]$_\alpha$, [0 1 0]$_\gamma$ to [$\bar{1}$ 1 0]$_\alpha$ and [0 0 1]$_\gamma$ to [0 0 1]$_\alpha$, by the Bain strain (Fig. 1), the correspondence matrix is simply:

$$(\alpha \ C \ \gamma) = \begin{pmatrix} 1 & \bar{1} & 0 \\ 1 & 1 & 0 \\ 0 & 0 & 1 \end{pmatrix} \qquad .....(20a)$$

If $\underline{u}$ is a vector defining the position of an atom in the γ unit cell, then it can be verified that (α C γ)[$\gamma;\underline{u}$] always gives a corresponding vector in the α lattice which terminates at a lattice point. For example, (1/2)[1 0 1]$_\gamma$ corresponds to (1/2)[1 1 1]$_\alpha$; both these vectors connect the origins of their respective unit cells to an atomic position. The Bain correspondence thus defines the position of each and every atom in the α lattice relative to the γ lattice. It is only possible to obtain a correspondence matrix like this when the primitive cells of each of the lattices concerned contain just one atom [5].

The primitive cell of the HCP lattice contains two atoms and any lattice correspondence will only define the final positions of an integral fraction of the atoms, the remainder having to shuffle into their correct positions in the product lattice. This can be demonstrated with the correspondence matrix for the example presented above. It is convenient to represent the conventional HCP lattice (basis H) in an alternative orthorhombic basis (symbol O), with basis vectors:

$$[1 \ 0 \ 0]_O = (1/2)[0 \ \bar{1} \ 1]_\gamma = [1 \ 0 \ \bar{1} \ 0]_H$$
$$[0 \ 1 \ 0]_O = (1/2)[2 \ \bar{1} \ \bar{1}]_\gamma = [\bar{1} \ \bar{2} \ \bar{1} \ 0]_H$$
$$[0 \ 0 \ 1]_O = (2/3)[1 \ 1 \ 1]_\gamma = [0 \ 0 \ 0 \ 1]_H$$

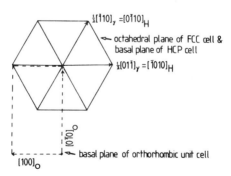

Fig. 17   Representation of Bases O, H, and γ

The orthorhombic unit cell thus contains three close-packed layers of atoms parallel to its (0 0 1) faces. The middle layer has atoms located at [0 1/3 1/2]$_O$, [1 1/3 1/2]$_O$ and [1/2 5/6 1/2]$_O$. The other two layers have atoms located at each corner of the unit cell and in the middle of each (0 0 1) face, as illustrated in Fig. 15.

From our earlier definition of a correspondence matrix, (O C γ) can be written directly from the relations (between basis vectors) stated earlier:

$$(\gamma \ C \ O) = (1/2) \begin{pmatrix} 0 & 2 & 1 \\ -1 & -1 & 1 \\ 1 & -1 & 2 \end{pmatrix}$$

Alternatively, the correspondence matrix (O C γ) may be derived (using eq.15) as follows:

$$(O \ C \ \gamma) = (O \ J \ \gamma)(\gamma \ P \ \gamma)$$

The matrix (γ P γ) is the total strain, which transforms the FCC lattice into the HCP lattice; it is equal to the matrix (Z P Z) derived in example 16, since the basis vectors of the orthonormal basis Z are parallel to the corresponding basis vectors of the orthogonal basis γ. It follows that:

$$\text{or } (O \ C \ \gamma) = \begin{pmatrix} 0 & -1 & 1 \\ 2/3 & -1/3 & -1/3 \\ 1/2 & 1/2 & 1/3 \end{pmatrix} \begin{pmatrix} 13/12 & 1/12 & 1/12 \\ 1/12 & 13/12 & 1/12 \\ -2/12 & -2/12 & 10/12 \end{pmatrix}$$

$$= \begin{pmatrix} -1/4 & -5/4 & 3/4 \\ 3/4 & -1/4 & -1/4 \\ 1/2 & 1/2 & 1/2 \end{pmatrix}$$

and

$$(\gamma \ C \ O) = (1/2) \begin{pmatrix} 0 & 2 & 1 \\ -1 & -1 & 1 \\ 1 & -1 & 2 \end{pmatrix}$$

Using this correspondence matrix, we can show that all the atoms, except those in the middle close-packed layer in the unit cell, have their positions relative to the parent lattice defined by the correspondence matrix. For example, the atom at the position $[1 \ 0 \ 0]_O$ corresponds directly to that at $[0 \ \bar{1} \ 1]_\gamma$ in the FCC lattice. However, $[0 \ 1/3 \ 1/2]_O$ corresponds to $(1/2)[7 \ 1 \ 4]_\gamma$ and there is no atom located at these co-ordinates in the γ lattice. The generation of the middle layer thus involves shuffles of $(1/12)[1 \ 1 \ \bar{2}]_\gamma$ as discussed earlier; we note that $(1/12)[7 \ 1 \ 4]_\gamma$ $(1/12)[1 \ 1 \ \bar{2}]_\gamma = (1/2)[1 \ 0 \ 1]_\gamma$ Thus, the atom at $[0 \ 1/3 \ 1/2]_O$ is derived from that at $(1/2)[1 \ 0 \ 1]_\gamma$ in addition to a shuffle displacement through $(a/12)[1 \ 1 \ \bar{2}]_\gamma$

The Conjugate of an Invariant-Plane Strain

We have already seen that an FCC lattice can be transformed to an HCP lattice by shearing the former on the system $\{1 \ 1 \ 1\}<1 \ 1 \ \bar{2}>$, $s = 8^{-1/2}$. This shear represents an invariant-plane strain (Z P Z) which can be factorised into a pure strain (Z Q Z) and a rigid body rotation (Z J Z), as in example 16. The pure deformation (Z Q Z) accomplishes the required lattice change from FCC to HCP, but is not an invariant-plane strain. As illustrated in Fig. 6 and in example 16, it is the rigid body rotation of 10.03° about $<1 \ \bar{1} \ 0>$ that makes the $\{1 \ 1 \ 1\}$ plane invariant and in combination with (Z Q Z) produces the final orientation relation implied by (Z P Z).

Referring to Fig. 6a,b, we see that there are in fact two ways [27] in which (Z Q Z) can be converted into an invariant-plane strain which transforms the FCC lattice to the HCP lattice. The first involves the rigid body rotation (Z J Z) in which c'd' is brought into coincidence with cd, as shown in Fig. 6c. The alternative would be to employ a rigid body rotation (Z J2 Z), involving a rotation of 10.03° about $<\bar{1} \ 1 \ 0>$, which would bring a'b' into coincidence with ab, making ab the trace of the invariant-plane. Hence, (Z Q Z) when combined with (Z J2 Z) would result in a different invariant-plane strain (Z P2 Z) which also shears the FCC lattice to the HCP lattice. From eq.8c,

46

(Z J2 Z) is given by:

$$(Z \ J2 \ Z) = \begin{pmatrix} 0.992365 & -0.007635 & -0.123091 \\ -0.007635 & 0.992365 & -0.123091 \\ 0.123092 & 0.123092 & 0.984732 \end{pmatrix}$$

From example 16, (Z Q Z) is given by:

$$(Z \ Q \ Z) = \begin{pmatrix} 1.094944 & 0.094943 & -0.020515 \\ 0.094943 & 1.094944 & -0.020515 \\ -0.020515 & -0.020515 & 0.841125 \end{pmatrix}$$

From eq.19a, (Z P2 Z) = (Z J2 Z)(Z Q Z)

$$\text{or} \quad (Z \ P2 \ Z) = \begin{pmatrix} 1.0883834 & 0.088384 & -0.123737 \\ 0.088384 & 1.088384 & -0.123737 \\ 0.126263 & 0.126263 & 0.823232 \end{pmatrix}$$

On comparing this with eq.11d, we see that (Z P2 Z) involves a shear of magnitude $s = 8^{-1/2}$ on $\{5 \ 5 \ \overline{7}\}_Z <7 \ 7 \ 10>_Z$. It follows that there are two ways of accomplishing the FCC to HCP change:

Mode 1:   Shear on $\{5 \ 5 \ \overline{7}\}_Z <7 \ 7 \ 10>_Z$ $s = 8^{-1/2}$

Mode 2:   Shear on $\{1 \ 1 \ 1\}_Z <1 \ 1 \ \overline{2}>_Z$ $s = 8^{-1/2}$

Both shears can generate a fully coherent interface between the FCC and HCP lattices (the coherent interface plane being coincident with the invariant-plane). Of course, while the $\{1 \ 1 \ 1\}$ interface of mode 2 would be atomically flat, the $\{5 \ 5 \ \overline{7}\}$ interface of mode 1 must probably be stepped on an atomic scale. The orientation relations between the FCC and HCP lattices would be different for the two mechanisms. In fact, (Z J2 Z) is

$$(Z \ J2 \ Z) = \begin{pmatrix} -0.2121216 & -1.2121216 & 0.6969703 \\ 0.7373739 & -0.2626261 & -0.2323234 \\ 0.3484848 & 0.3484848 & 0.7121212 \end{pmatrix}$$

It is intriguing that only the second mode has been observed experimentally, even though both involve identical shear magnitudes.

A general conclusion to be drawn from the above analysis is that whenever two lattices can be related by an IPS (i.e., whenever they can be joined by a fully coherent interface), it is always possible to find a *conjugate* IPS which in general allows the two lattices to be differently oriented but still connected by a fully coherent interface. This is clear from Fig. 6 where we see that there are two ways of carrying out the rigid body rotation in order to obtain an IPS which transforms the FCC lattice to the HCP lattice. The deformation involved in twinning is also an IPS so that for a given twin mode it ought to be possible to find a conjugate twin mode. In Fig. 13b, a rigid body rotation about $[\overline{1} \ 1 \ 0]$, which brings K_2 into coincidence with K_2' would give the conjugate twin mode on $(1 \ 1 \ \overline{1})[1 \ 1 \ 2]$.

We have used the pure strain (Z Q Z) to transform the FCC crystal into a HCP crystal. However, before this transformation, we could use any of an infinite number of operations (e.g., a symmetry operation) to bring the FCC lattice into self-coincidence. Combining any one of these operations with (Z Q Z) then gives us an alternative deformation which can accomplish the FCC→HCP lattice change without altering the orientation relationship. It follows that two lattices can be deformed into one another in an infinite number of ways. Hence, *prediction* of the transformation strain is not possible in the sense that intuition or experimental evidence has to be used to choose the 'best' or 'physically most meaningful' transformation strain.

Example 17: The Combined Effect of two Invariant-Plane Strains

Show that the combined effect of the operation of two arbitrary invariant-plane strains is equivalent to an invariant-line strain (ILS). Hence prove that if the two invariant-plane strains have the same invariant-plane, or the same displacement direction, then their combined effect is simply another IPS [2].

The two invariant plane strains are referred to an orthonormal basis X and are designated (X P X) and (X Q X), such that m and n are their respective magnitudes, $\underline{d}$ and $\underline{e}$ their respective unit displacement directions and $\underline{p}$ and $\underline{q}$ their respective unit invariant-plane normals. If (X Q X) operates first, then the combined effect of the two strains (eq.11e) is

$$(X\ P\ X)(X\ Q\ X) = \{I + m[X;\underline{d}](\underline{p};X^*)\}\{I + n[X;\underline{e}](\underline{q};X^*)\}$$
$$= I + m[X;\underline{d}](\underline{p};X^*) + n[X;\underline{e}](\underline{q};X^*) + mn[X;\underline{d}](\underline{p};X^*)[X;\underline{e}](\underline{q};X^*)$$
$$= I + m[X;\underline{d}](\underline{p};X^*) + n[X;\underline{e}](\underline{q};X^*) + g[X;\underline{d}](\underline{q};X^*) \qquad .....(21a)$$

where g is the scalar quantity $g = mn(\underline{p};X^*)[X;\underline{e}]$

If $\underline{u}$ is a vector which lies in both the planes represented by $\underline{p}$ and $\underline{q}$, i.e., it is parallel to $\underline{p}\wedge\underline{q}$, then it is obvious (eq.21a) that $(X\ P\ X)(X\ Q\ X)[X;\underline{u}] = [X;\underline{u}]$, since $(\underline{p};X^*)[X;\underline{u}] = 0$ and $(\underline{q};X^*)[X;\underline{u}] = 0$. It follows that $\underline{u}$ is parallel to the invariant line of the total deformation (X S X) = (X P X)(X Q X). This is logical since (X P X) should leave every line on $\underline{p}$ invariant and (X Q X) should leave all lines on $\underline{q}$ invariant. The line that is common to both $\underline{p}$ and $\underline{q}$ should therefore be unaffected by (X P X)(X Q X), as is clear from eq.21a. Hence, the combination of two arbitrary invariant-plane strains (X P X)(X Q X) gives an *Invariant-Line Strain* (X S X).

If $\underline{d} = \underline{e}$, then from eq.21a

$$(X\ P\ X)(X\ Q\ X) = I + [X;\underline{d}](\underline{r};X^*)$$

where $(\underline{r};X^*) = m(\underline{p};X^*) + n(\underline{q};X^*) + g(\underline{q};X^*)$

which is simply another IPS on a plane whose normal is parallel to $\underline{r}$. If $\underline{p} = \underline{q}$, then from eq.21a

$$(X\ P\ X)(X\ Q\ X) = I + [X;\underline{f}](\underline{p};X^*)$$

where $[X;\underline{f}] = m[X;\underline{d}] + n[X;\underline{e}] + g[X;\underline{d}]$

which is an IPS with a displacement direction parallel to $[X;\underline{f}]$.

Hence, in the special case where the two IPSs have their displacement directions parallel, or have their invariant-plane normals parallel, their combined effect is simply another IPS.

It is interesting to examine how plane normals are affected by invariant-line strains. Taking the inverse of (X S X), we see that

$$(X\ S\ X)^{-1} = (X\ Q\ X)^{-1}(X\ P\ X)^{-1}$$

or from eq.13,

$$(X\ S\ X)^{-1} = \{I - an[X;\underline{e}](\underline{q};X^*)\}\{I - bm[X;\underline{d}](\underline{p};X^*)\}$$

$$= I - an[X;\underline{e}](\underline{q};X^*) - bm[X;\underline{d}](\underline{p};X^*) + cnm[X;\underline{e}](\underline{p};X^*)$$

$$.....(21b)$$

where a, b and c are scalar constants given by

$1/a = \det(X\ Q\ X)$, $1/b = \det(X\ P\ X)$ and $c = ab(\underline{q};X^*)[X;\underline{d}]$.

If $\underline{h} = \underline{e}\wedge\underline{d}$, then $\underline{h}$ is a reciprocal lattice vector representing the plane which contains both $\underline{e}$ and $\underline{d}$. It is evident from eq.21b that $(\underline{h};X^*)(X\ S\ X)^{-1} = (\underline{h};X^*)$, since $(\underline{h};X^*)[X;\underline{e}] = 0$ and $(\underline{h};X^*)[X;\underline{d}] = 0$. In other words, the plane normal $\underline{h}$ is an invariant normal of the invariant-line strain $(X\ S\ X)^{-1}$.

We have found that an ILS has two important characteristics: it leaves a line $\underline{u}$ invariant and also leaves a plane normal $\underline{h}$ invariant. If the ILS is factorised into two IPS's, then $\underline{u}$ lies at the intersection of the invariant-planes of these component IPS's, and $\underline{h}$ defines the plane containing the two displacement vectors of these IPS's. These results will be useful in understanding martensite.

MARTENSITIC TRANSFORMATIONS

In this chapter we develop a fuller description of martensitic transformations, focussing attention on steels, although the concepts involved are applicable to materials as diverse as A15 superconducting compounds [35] and Ar-N_2 solid solutions [36]. The fundamental requirement for martensitic transformation is that the shape deformation accompanying diffusionless transformation be an invariant-plane strain; all the characteristics of martensite will be shown to be consistent with this condition. In this chapter, we refer to martensite in general as α' and body-centered cubic martensite as α.

The Diffusionless Nature of Martensitic Transformations

Diffusion means the 'mixing up of things'; martensitic transformations are by definition [37] diffusionless. The formation of martensite *can* occur at very low temperatures where atomic mobility may be inconceivably small. The diffusion, even of atoms in interstitial sites, is then not possible within the time scale of the transformation. The martensite-start temperature (M_s) is the highest temperature at which martensite forms on cooling the parent phase. Some examples of M_s temperatures are given below:

Table 2: Martensite-Start Temperatures

Composition, wt.pct.	M_s, K
Fe-31.0Ni-0.23C	83
Fe-33.5Ni-0.22C	<4
Fe-3.0Mn-2.0Si-0.4C	493
Cu-15Al	253
Ar-40N_2	30

Even when martensite forms at high temperatures, its rate of growth can be so high that diffusion does not occur. Plates of martensite in iron based alloys are known to grow at speeds approaching that of sound in the metal [38,39]; such speeds are generally inconsistent with diffusion occurring during transformation. Furthermore, the composition of martensite can be measured and shown to be identical to that of the parent phase (although this in itself does not constitute evidence for diffusionless transformation).

The Interface between the Parent and Product Phases

The fact that martensite can form at very low temperatures also means that any process which is a part of its formation process cannot rely on thermal activation. For instance, the interface connecting the martensite with the parent phase must be able to move easily at very low temperatures, without any significant help from thermal agitation (throughout this text, the terms interface and interface plane refer to the average interface, as determined on a macroscopic scale). Because the interface must have high mobility at low temperatures and at high velocities, it cannot be incoherent; it must therefore be semi-coherent or fully coherent [40]. Fully coherent interfaces are of course only possible when the parent and product lattices can be related by a strain which is an invariant-plane strain [5]. In the context of martensite, we are concerned with interphase-interfaces

51

and fully coherent interfaces of this kind are rare for particles of appreciable size; the FCC→HCP transformation is one example where a fully coherent interface is possible. Martensitic transformation in ordered Fe_3Be occurs by a simple shearing of the lattice (an IPS) [41], so that a fully coherent interface is again possible. More generally, the interfaces tend to be semi-coherent. For example, it was discussed in chapter 1 that a FCC austenite lattice cannot be transformed into a BCC martensite lattice by a strain which is an IPS, so that these lattices can be expected to be joined by semi-coherent interfaces.

The semi-coherent interface should consist of coherent regions separated periodically by discontinuities which prevent the misfit in the interface plane from accumulating over large distances, in order to minimise the elastic strains associated with the interface. There are two kinds of semi-coherency [5,27]; if the discontinuities mentioned above are intrinsic dislocations with Burgers vectors in the interface plane, not parallel to the dislocation line, then the interface is said to be epitaxially semi-coherent. The term 'intrinsic' means that the dislocations are a necessary part of the interface structure and have not simply strayed into the boundary - they do not have a long-range strain field. The normal displacement of such an interface requires the thermally activated climb of intrinsic dislocations, so that the interface can only move in a non-conservative manner, with relatively restricted or zero mobility at low temperatures. A martensite interface cannot therefore be epitaxially semi-coherent.

In the second type of semi-coherency, the discontinuities discussed above are screw dislocations, or dislocations whose Burgers vectors do not lie in the interface plane. This kind of semi-coherency is of the type associated with glissile martensite interfaces, whose motion is conservative (i.e., the motion does not lead to the creation or destruction of lattice sites). Such an interface should have a high mobility since the migration of atoms is not necessary for its movement. Actually, two further conditions must be satisfied before even this interface can be said to be glissile:

1) A glissile interface requires that the glide planes of the intrinsic dislocations associated with the product lattice must meet the corresponding glide planes of the parent lattice edge to edge in the interface [27], along the dislocation lines.

2) If more than one set of intrinsic dislocations exist, then these should either have the same line vector in the interface, or their respective Burgers vectors must be parallel [27]. This condition ensures that the interface can move as an integral unit. It also implies (example 17) that the deformation caused by the intrinsic dislocations, when the interface moves, can always be described as a simple shear (caused by a resultant intrinsic dislocation which is a combination of all the intrinsic dislocations) on some plane which makes a finite angle with the interface plane, and intersects the latter along the line vector of the resultant intrinsic dislocation.

Obviously, if the intrinsic dislocation structure consists of just a single set of parallel dislocations, or of a set of different dislocations which can be summed to give a single glissile intrinsic dislocation, then it follows that there must exist in the interface, a line which is parallel to the resultant intrinsic dislocation line vector, along which there is zero distortion. Because this line exists in the interface, it is also unrotated. It is an *invariant-line* in the interface between the parent and product lattices. When full coherency between the parent and martensite lattices is not possible, then for the interface to be glissile, the transformation strain relating the two lattices must be an invariant-line strain, with the invariant-line being in the interface plane.

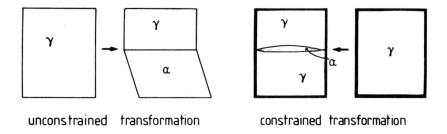

unconstrained transformation   constrained transformation

Fig. 18: The habit plane of martensite (α') under conditions of unconstrained and constrained transformation, respectively. In the latter case, the dashed line indicates the trace of the habit plane.

The interface between the martensite and the parent phase is usually called the 'habit plane'; when the transformation occurs without any constraint, the habit plane is macroscopically flat, as illustrated in Fig. 18. When the martensite forms in a constrained environment, it grows in the shape of a thin lenticular plate or lath and the habit plane is a little less clear in the sense that the interface is curved on a macroscopic scale. However, it is experimentally found that the average plane of the plate (the plane containing the major circumference of the lens) corresponds closely to that expected from crystallographic theory, and to that determined under conditions of unconstrained transformation. The aspect ratio (maximum thickness to length ratio) of lenticular plates is usually less than 0.05, so that the interface plane does not depart very much from the average plane of the plate. Some examples of habit plane indices (relative to the austenite lattice) are given below:

Table 3: Habit Planes of Martensite. The quoted indices are approximate, since the habit planes are in general irrational.

Composition, wt.pct.	Approximate Habit Plane Indices
Low-alloy steels, Fe-28Ni	{1 1 1}
Plate martensite in Fe-1.8C	{2 9 5}
Fe-30Ni-0.3C	{3 15 10}
Fe-8Cr-1C	{2 5 2}
ε martensite in 18/8 stainless steel	{1 1 1}

Orientation Relationships

Since there is no diffusion during martensitic transformation, atoms must be transferred across the interface in a co-ordinated manner (a "military transformation" - Ref.42) and it follows that the austenite and martensite lattices should be intimately related. All martensite transformations lead to a reproducible orientation relationship between the parent and product lattices. The orientation relationship usually consists of parallel or very nearly parallel corresponding closest-packed planes from the two lattices, and it is usually the case that the corresponding close-packed directions in these planes are also roughly parallel. Typical examples of orientation relations found in steels are given below; these are stated in a simple manner for illustration purposes although the best way of specifying orientation relations is by the use of co-ordinate transformation matrices, as in chapter 2.

Kurdjumov-Sachs [15]

$$\{1\ 1\ 1\}_{\gamma} \| \{0\ 1\ 1\}_{\alpha'}$$
$$<1\ 0\ \bar{1}>_{\gamma} \| <1\ 1\ \bar{1}>_{\alpha'}$$

Nishiyama-Wasserman [16]

$$\{1\ 1\ 1\}_{\gamma} \| \{0\ 1\ 1\}_{\alpha'}$$

$$<1\ 0\ \bar{1}>_{\gamma} \quad \text{about } 5^{0} \quad \text{from } <1\ 1\ \bar{1}>_{\alpha'} \quad \text{towards } <\bar{1}\ 1\ \bar{1}>_{\alpha'}$$

$$<1\ 1\ \bar{2}>_{\gamma} \| <0\ \bar{1}\ 1>_{\alpha'}$$

Greninger-Troiano [43]

$$\{1\ 1\ 1\}_{\gamma} \quad 0.2^{0} \quad \text{from } \{0\ 1\ 1\}_{\alpha'}$$
$$<1\ 0\ 1>_{\gamma} \quad 2.7^{0} \quad \text{from } <1\ 1\ \bar{1}>_{\alpha'}$$

The electron diffraction pattern shown in Fig. 19, taken to include both γ and α, indicates how well the two lattices are "matched" in terms of the orientation relation, even though the lattice types are different. The reflections from austenite are identified by the subscript 'a'.

In studying martensitic transformations in steels, one of the major conceptual difficulties is to explain why the observed orientation relations differ from that implied by the Bain strain. The previous section on martensite interfaces, and the results of chapter one explain this 'anomaly'. As we have already seen, the martensite interface must contain an invariant-line, and the latter can only be obtained by combining the Bain Strain with a rigid body rotation. This combined set of operations amounts to the necessary invariant-line strain and the rigid body rotation component changes the orientation between the parent and product lattices to the experimentally observed relation.

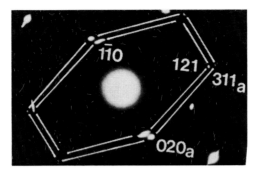

Fig. 19: Electron diffraction pattern from BCC martensite and FCC austenite lattices in steels (the austenite reflections are identified by the subscript 'a').

The Shape Deformation due to Martensitic Transformation

All martensitic transformations involve co-ordinated movements of atoms and are diffusionless. Since the shape of the pattern in which the atoms in the parent crystal are arranged nevertheless changes in a way that is consistent with the change in crystal structure on martensitic transformation, it follows that there must be a physical change in the macroscopic shape of the parent crystal during transformation [44]. The shape deformation and its significance can best be illustrated by reference to Fig. 20, where a comparison is made between diffusional and diffusionless transformations. For simplicity, the diagram refers to a case where the transformation strain is an invariant-plane strain and a fully coherent interface exists between the parent and product lattices, irrespective of the mechanism of transformation.

Considering the shear transformation first, we note that since the pattern of atomic arrangement is changed on transformation, and since the transformation is diffusionless, the macroscopic shape of the crystal changes. The shape deformation has the exact characteristics of an IPS. The initially flat surface normal to da becomes tilted about the line formed by the intersection of the interface plane with the surface normal to da. The straight line ab is bent into two connected and straight segments ae and eb. Hence, an observer looking at a scratch that is initially along ab and in the surface abcd would note that on martensite formation, the scratch becomes homogeneously deflected about the point e where it intersects the trace of the interface plane. Furthermore, the scratches ae and eb would be seen to remain connected at the point e. This amounts to proof that the shape deformation has, on a macroscopic scale, the characteristics of an IPS and that the interface between the parent and product lattices does not contain any distortions (i.e., it is an invariant-plane). Observing the deflection of scratches is one way of experimentally deducing the nature of shape deformations accompanying transformations.

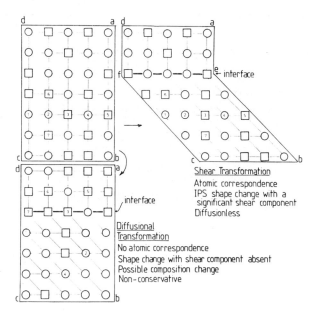

Shear Transformation
Atomic correspondence
IPS shape change with a
significant shear component
Diffusionless

Diffusional
Transformation
No atomic correspondence
Shape change with shear component absent
Possible composition change
Non-conservative

<u>Fig. 20</u>:   Schematic illustration of the mechanisms of diffusional and shear transformations. The lines connect corresponding directions.

In Fig. 20 it is also implied that the martensitic transformation is diffusionless; the labelled rows of atoms in the parent crystal remain in the correct sequence in the martensite lattice, despite transformation. Furthermore, it is possible to suggest that a particular atom in the martensite must have originated from a corresponding particular atom in the parent crystal (e.g., atom 3 in the martensite can be uniquely identified with site 3 in the parent, since there is no mixing up of the atoms during transformation). A formal way of expressing this property is to say that there exists an *atomic correspondence* between the parent and product lattices.

In the case of the diffusional transformation illustrated in Fig. 20, it is evident that the product phase can be of a different composition from the parent phase. In addition, there has been much mixing up of atoms during transformation and the order of arrangement of atoms in the product lattice is different from that in the parent lattice - the atomic correspondence has been destroyed (atom no.3 has wandered away from its corresponding site in the product). It is no longer possible to suggest that a particular atom in the product phase originates from a certain site in the parent lattice. Because the transformation involves a reconstruction of the parent lattice, atoms are able to diffuse around in such a way that the IPS shape deformation (and its accompanying strain energy) does not arise. The scratch ab remains straight across the interface and is unaffected by the transformation.

In summary, martensitic transformations are always accompanied by a change in shape of the parent crystal; this shape deformation always has the characteristics of an invariant-plane strain, when examined on a macroscopic scale. The occurrence of such a shape deformation is taken to imply the existence of an atomic correspondence between the parent and product lattices. It is possible to state that a particular atom in the product occupied a particular corresponding site in the parent lattice.

These results have some interesting consequences. The formation of martensite in a constrained environment must (due to its IPS shape deformation) cause a distortion of the parent lattice in its vicinity. The strain energy due to this distortion, per unit volume of martensite, is approximately given by [28-30] $E=(c/r)\mu(s^2 + \delta^2)$ where μ is the shear modulus of the parent lattice, (c/r) is the thickness to length ratio of the martensite plate and s and δ are the shear and dilatational components of the shape deformation strain. It follows that martensite must always have a thin plate morphology, if E is to be minimised and this is of course experimentally found to be the case. E usually amounts to about 600Jmol^{-1} for martensite in steels [30], when the shape deformation is entirely elastically accommodated. If the austenite is soft, then some plastic accommodation may occur but the E value calculated on the basis of purely elastic accommodation should be taken to be the upper limit of the stored energy due to the shape change accompanying martensitic transformation. This is because the plastic accommodation is driven by the shape deformation and it can only serve to mitigate the effects of the shape change [30]. In the event that plastic accommodation occurs, dislocations and other defects may be generated both in the parent and product lattices.

The Phenomenological Theory of Martensite Crystallography

We have emphasized that a major feature of the martensite transformation is its shape deformation, which on a macroscopic scale has the characteristics of an invariant-plane strain. The magnitude m of the shape deformation can be determined as can its unit displacement vector d. The habit plane of the martensite (unit normal p) is the invariant-plane of the shape deformation. The shape deformation can be represented by means of a shape deformation matrix (F P F) such that:

$$(F\ P\ F) = I + m[F;\underline{d}](\underline{p};F^*)$$

where the basis F is for convenience chosen to be orthonormal, although the equation is valid for any basis.

For the shear transformation illustrated in Fig. 20 and for the FCC→HCP martensite reaction, the lattice transformation strain is itself an IPS and there is no difficulty in reconciling the transformation strain and the observed shape deformation. In other words, if the parent lattice is operated on by the shape deformation matrix, then the correct product lattice is generated if shuffles are allowed; the transformation strain is the same as the shape deformation. This is not the case [45] for the FCC→BCC martensite reaction and for many other martensite transformations where the lattice transformation strain (F S F) does not equal the observed shape deformation (F P F). In example 2 it was found that the Bain Strain (F B F) when combined with an appropriate rigid body rotation (F J F) gives an invariant-line strain which when applied to the FCC lattice generates the BCC martensite lattice. However, the shape deformation that accompanies the formation of BCC martensite from austenite is nevertheless experimentally found to be an invariant-plane strain. This is the major anomaly that the theory of martensite crystallography attempts to resolve: the experimentally observed shape deformation is inconsistent with the lattice transformation strain. If the observed shape deformation is applied to the parent lattice then the austenite lattice is deformed into an intermediate lattice (not experimentally observed) but not into the required BCC lattice.

This anomaly is schematically illustrated in Figs. 20a-c. Fig. 21a represents the shape of the starting austenite crystal with the FCC structure. On martensitic transformation its shape alters to that illustrated in Fig. 21b and the shape deformation on going from (a) to (b) is clearly an IPS on the plane with unit normal p and in the unit displacement direction d. However, the structure of the crystal in Fig. 21b is some intermediate lattice which is not BCC, since an IPS cannot on its own change the FCC structure to the BCC structure. An invariant-line strain can however transform FCC to BCC, and since an ILS can be factorised into two invariant-plane strains, it follows that the further deformation (F Q F) needed to change the intermediate structure of Fig. 21b to the BCC structure (Fig. 21c) is another IPS. If the deformation (F Q F) is of magnitude n on a plane with unit normal q and in a unit direction $\underline{e}$, then:

$$(F \ Q \ F) = I + n[F;\underline{e}](q;F^*)$$

(F Q F) has to be chosen in such a way that (F P F)(F Q F) = (F S F), where (F S F) is an invariant-line strain which transforms the FCC lattice to the BCC lattice. Hence, a combination of two invariant-plane strains can accomplish the necessary lattice change but this then gives the wrong shape change as the extra shape change due to (F Q F), in changing (b) to (c), is not observed.

Experiments [46-48] indicate that the shape deformation due to the FCC→BCC martensite transformation is an IPS, and it seems that the effect of (F Q F) on the macroscopic shape is invisible. If we can find a way of making the effect of (F Q F) invisible as far as the shape change is concerned, then the problem is essentially determined.

(F Q F) can be made invisible by applying another deformation to (c) such that the shape of (c) is brought back to that of (b), without altering the BCC structure of (c). Such a deformation must therefore be *lattice-invariant* because it must not alter the symmetry or unit cell dimensions of the parent crystal structure. Ordinary slip does not change the nature of the lattice and is one form of a lattice-invariant deformation. Hence, slip deformation on the planes q and in the direction -$\underline{e}$ would make the shape change due to (F Q F) invisible on a macroscopic scale, as illustrated in Fig. 21d. The magnitude of this lattice-invariant slip shear is of course determined by that of (F Q F) and we know that it is not possible to continuously vary the magnitude of slip shear, since the Burgers vectors of slip dislocations are discrete. (F Q F) on the other hand can have any arbitrary magnitude. This difficulty can be overcome by applying the slip shear *inhomogeneously*, by the passage of a discrete slip dislocation on say every nth plane, which has the effect of allowing the magnitude of the lattice-invariant shear to vary as a function of n. In applying the lattice-invariant shear to (c) in order to obtain (d), the BCC structure of (c) is completely unaffected, while is shape is deformed inhomogeneously to correspond to that of (b), as illustrated in Fig. 21d.

This then is the essence of the theory of martensite crystallography [1,2,4], which explains the contradiction that the lattice transformation strain is an ILS but the macroscopic shape deformation is an IPS. The lattice transformation strain when combined with an inhomogeneous lattice-invariant shear produces a macroscopic shape change which is an IPS.

Twinning is another deformation which does not change the nature of the lattice (although unlike slip, it reorientates it); the shape (c) of Fig. 21 could be deformed to correspond macroscopically to that of (b), without changing its BCC nature, by twinning, as illustrated in Fig. 21e. The magnitude of the lattice-invariant deformation can be adjusted by varying the volume fraction of the twin. This explains the twin substructure found in many ferrous martensites, and such twins are called transformation twins. The irrationality of the habit planes arises because the indices of the habit plane depend on the amount of lattice-invariant deformation, a quantity which does not necessarily correlate with displacements equal to discrete lattice vectors.

Since the effect of the homogeneous strain (F Q F) on the shape of the parent crystal has to be cancelled by another opposite but inhomogeneously applied lattice-invariant deformation, it follows that (F Q F) is restricted to being a simple shear with the displacement vector $\underline{e}$ being confined to

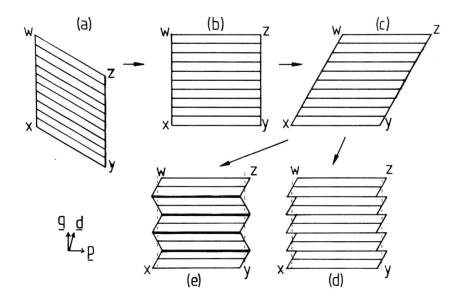

<u>Fig. 21:</u>   Schematic illustration of the phenomenological theory of martensite. (a) represents the austenite crystal and (c), (d) & (e) all have a BCC structure. (b) has a structure between FCC and BCC, $\underline{p}$ is the habit plane unit normal and $\underline{q}$ is the unit normal to the plane on which the lattice-invariant shear occurs. The heavy horizontal lines in (e) are coherent twin boundaries. Note that the vector $\underline{e}$ is normal to $\underline{q}$ but does not lie in the plane of the diagram.

the invariant-plane of (F Q F). In other words, (F Q F) must have a zero dilatational component, since it cannot otherwise be cancelled by another IPS which preserves the lattice [2]. As noted earlier, lattice-invariant deformations cannot alter the volume or symmetry of the lattice. The determinant of a deformation matrix gives the ratio of the volume after deformation to that prior to deformation, so that det (F Q F)=1. This means that the total volume change of transformation is given by det (F P F) = det (F S F).

In summary, the martensite transformation in iron requires an invariant-line strain (F S F) to change the FCC lattice to the BCC martensite lattice and to obtain the experimentally observed orientation relation. This can be imagined to consist of two homogeneous invariant-plane strains (F P F) and (F Q F), such that (F S F) = (F P F)(F Q F). However, the shape change due to the simple shear (F Q F) is rendered invisible on a macroscopic scale since there is also an inhomogeneous lattice-invariant deformation (which can be slip or twinning) which cancels out the shape change due to (F Q F), without altering the lattice structure. It follows that the macroscopic shape change observed is solely due to (F P F) and therefore has the characteristics of an invariant-plane strain, as experimentally observed. We have already seen that the transformation strain (F S F) can be factorised into a Bain Strain (F B F) combined with an appropriate rigid body rotation (F J F), such that (F S F)=(F J F)(F B F) and is an invariant-line strain, with the invariant-line lying in the planes p and q, and the invariant-normal of (F S F) defining a plane containing d and e. Hence, the theory of martensite can be summarised in terms of the equation

$$(F \ S \ F)=(F \ J \ F)(F \ B \ F)=(F \ P \ F)(F \ Q \ F) \qquad .....(22a)$$

Stage 1: Calculation of the Lattice Transformation Strain

Two arbitrary lattices can be transformed into one another by an infinite number of different transformation strains, but only some of these may have reasonably small principal deformations. The choice available can be further reduced by considering only those strains which involve the minimum degree of shuffling of atoms and by considering the physical implications of such strains. In the case of martensitic transformations, a further condition has to be satisfied; the lattice transformation strain must also be an invariant-line strain if the interface is to be glissile [27].

For the FCC→BCC martensitic transformation, the Bain Strain, which is a pure deformation, involves the smallest atomic displacements during transformation. When it is combined with an appropriate rigid body rotation (example 4), the total strain amounts to an invariant-line strain. For martensitic transformations, the rigid body rotation has to be chosen in such a way that the invariant-line lies in the plane of the lattice-invariant shear and also in the habit plane of the martensite; these planes are the invariant planes of (F Q F) and (F P F) respectively, so that the line common to these planes is not affected by these deformations. Furthermore, the invariant normal of the ILS must define a plane which contains the displacement directions of the lattice invariant shear and of the shape deformation. This ensures that the spacing of this plane is not affected by (F Q F) or (F P F).

Example 18 below illustrates how the transformation strain can be determined once the nature of the pure deformation (Bain Strain) which accomplishes the lattice change is deduced using the procedures discussed above. To ensure that the invariant-line and invariant-normal of the transformation strain are compatible with the mode of lattice-invariant shear, we first need to specify the latter. In example 18 it is assumed that the plane and direction of the lattice- invariant shear are $(1\ 0\ 1)_F$ and $[1\ 0\ \bar{1}]_F$ respectively. One variant of the Bain Strain is illustrated in Fig. 1, where we see that $[1\ 0\ 0]_\gamma$ is deformed into $[1\ 1\ 0]_\alpha$, $[0\ 1\ 0]_\gamma$ to $[\bar{1}\ 1\ 0]_\alpha$ and $[0\ 0\ 1]_\gamma$ to $[0\ 0\ 1]_\alpha$, so that the variant of the Bain correspondence matrix is given by eq.20a. We will use this variant of the Bain correspondence matrix throughout the text, but we note that there are two other possibilities, where $[0\ 0\ 1]_\alpha$ can be derived from either $[1\ 0\ 0]_\gamma$ or $[0\ 1\ 0]_\gamma$ respectively.

Example 18: Determination of the Lattice Transformation Strain

The deformation matrix representing the Bain Strain, which carries the FCC austenite lattice (Fig. 1) to the BCC martensite lattice is given by

$$(F\ B\ F)\ =\ \begin{pmatrix} \eta_1 & 0 & 0 \\ 0 & \eta_2 & 0 \\ 0 & 0 & \eta_3 \end{pmatrix}$$

where F is an orthonormal basis consisting of unit basis vectors (f_i) parallel to the crystallographic axes of the conventional FCC austenite unit cell (Fig. 1, $f_1\|a_1$, $f_2\|a_2$ & $f_3\|a_3$). η_i are the principal deformations of the Bain Strain, given by $\eta_1=\eta_2=1.136071$ and $\eta_3=0.803324$.

Find the rigid body rotation (F J F) which when combined with the Bain Strain gives an invariant-line strain (F S F)=(F J F)(F B F), subject to the condition that the invariant-line of (F S F) must lie in $(1\ 0\ 1)_F$ and that the plane defined by the invariant-normal of (F S F) contains $[1\ 0\ \bar{1}]_F$.

Writing the invariant-line as $[F;\underline{u}]=[u_1\ u_2\ u_3]$, we note that for $\underline{u}$ to lie in $(1\ 0\ 1)_F$, its components must satisfy the equation

$$u_1\ =\ -u_3 \qquad\qquad .....(23a)$$

Prior to deformation, $\qquad |\underline{u}|^2\ =\ (\underline{u};F)[F;\underline{u}]\ =\ 1 \qquad\qquad .....(23b)$

$\underline{u}$, as a result of deformation becomes a new vector $\underline{w}$ with

$$\begin{aligned} |\underline{w}|^2\ &=\ (\underline{w};F)[F;\underline{w}] \\ &=\ (\underline{u};F)(F\ B'\ F)(F\ B\ F)[F;\underline{u}] \\ &=\ (\underline{u};F)(F\ B\ F)^2[F;\underline{u}] \end{aligned}$$

if the magnitude of $\underline{u}$ is not to change on deformation then $|\underline{u}|\ =\ |\underline{w}|$ or

$$u_{\underline{1}}^2 + u_{\underline{2}}^2 + u_{\underline{3}}^2 = \eta_1^2 u_{\underline{1}}^2 + \eta_2^2 u_{\underline{2}}^2 + \eta_3^2 u_{\underline{3}}^2 \qquad \dots\dots(23c)$$

Equations 23a-c can be solved simultaneously to give <u>two</u> solutions for undistorted lines:
$$[F;\underline{u}] = [-0.671120 \quad -0.314952 \quad 0.671120]$$
$$[F;\underline{v}] = [-0.671120 \quad 0.314952 \quad 0.671120]$$

To solve for the invariant normal of the ILS, we proceed as follows: writing $(\underline{h};F^*)=(h_1 \quad h_2 \quad h_3)$, we note that for $\underline{h}$ to contain $[1 \ 0 \ \overline{1}]_F$, its components must satisfy the equation
$$h_1 = h_3 \qquad \dots\dots(23d)$$
Furthermore,
$$(\underline{h};F^*)[F^*;\underline{h}] = 1 \qquad \dots\dots(23e)$$
$\underline{h}$, on deformation becomes a new plane normal $\underline{l}$ and if $|\underline{h}|=|\underline{l}|$ then
$$|\underline{l}|^2 = (\underline{l};F^*)[F^*;\underline{l}]$$
$$= (\underline{h};F^*)(F \ B \ F)^{-1}(F \ B' \ F)^{-1}[F^*;\underline{h}]$$
so that
$$h_1^2 + h_2^2 + h_3^2 = (l_1/\eta_1)^2 + (l_2/\eta_2)^2 + (l_3/\eta_3)^2 \qquad \dots\dots(23f)$$

Solving eqs.23d-f simultaneously, we obtain the <u>two</u> possible solutions for
the undistorted-normals as
$$(\underline{h};F^*) = (0.539127 \quad 0.647058 \quad 0.539127)$$
$$(\underline{k};F^*) = (0.539127 \quad -0.647058 \quad 0.539127)$$
To convert $(F \ B \ F)$ into an invariant-line strain $(F \ S \ F)$ we have to employ a rigid body rotation $(F \ J \ F)$ which simultaneously brings an undistorted line (such as $\underline{w}$) and an undistorted normal (such as $\underline{l}$) back into their original directions along $\underline{u}$ and $\underline{h}$ respectively. This is possible because the angle between $\underline{w}$ and $\underline{l}$ is the same as that between $\underline{u}$ and $\underline{h}$, as shown below:
$$\underline{l}.\underline{w} = (\underline{l};F^*)[F;\underline{w}]$$
$$= (\underline{h};F^*)(F \ B \ F)^{-1}(F \ B \ F)[F;\underline{u}]$$

$$= (\underline{h};F^*)[F;\underline{u}]$$
$$= \underline{h}.\underline{u}$$

Hence, one way of converting $(F \ B \ F)$ into an ILS is to employ a rigid body rotation which simultaneously rotates $\underline{l}$ into $\underline{h}$ and $\underline{w}$ into $\underline{u}$. Of course, we have found that there are two undistorted lines and two undistorted normals which satisfy the conditions of the original question, and there are clearly four ways of choosing pairs of undistorted lines and undistorted normals (in the case we investigate, the four solutions are clearly crystallographically equivalent). There are therefore four solutions (different in general) to the problem of converting $(F \ B \ F)$ to $(F \ S \ F)$, subject to the condition that the invariant-line should be in $(1 \ 0 \ 1)$ and that the invariant normal defines a plane containing $[1 \ 0 \ \overline{1}]$. We will concentrate on the solution obtained using the pair $\underline{u}$ and $\underline{h}$.

Now,
$$\underline{l} = (\underline{h};F^*)(F \ B \ F)^{-1} = (0.474554 \quad 0.569558 \quad 0.671120)$$
and
$$\underline{w} = (F \ B \ F)[F;\underline{u}] = [-0.762440 \quad -0.357809 \quad 0.539127]$$
and
$$\underline{a} = \underline{u}/\underline{h} = (-0.604053 \quad 0.723638 \quad -0.264454)$$
and
$$\underline{b} = \underline{w}/\underline{l} = (-0.547197 \quad 0.767534 \quad -0.264454)$$

The required rigid body rotation should rotate $\underline{w}$ back to $\underline{u}$, $\underline{l}$ back to $\underline{h}$ and $\underline{b}$ to $\underline{a}$, giving the three equations:

$$[F;\underline{u}] = (F \ J \ F)[F;\underline{w}]$$
$$[F;\underline{h}] = (F \ J \ F)[F;\underline{l}]$$
$$[F;\underline{a}] = (F \ J \ F)[F;\underline{b}]$$

which can be expressed as a 3x3 matrix equation

$$\begin{pmatrix} u_1 & h_1 & a_1 \\ u_2 & h_2 & a_2 \\ u_3 & h_3 & a_3 \end{pmatrix} = \begin{pmatrix} J_{11} & J_{12} & J_{13} \\ J_{21} & J_{22} & J_{23} \\ J_{31} & J_{32} & J_{33} \end{pmatrix} \begin{pmatrix} w_1 & l_1 & b_1 \\ w_2 & l_2 & b_2 \\ w_3 & l_3 & b_3 \end{pmatrix}$$

It follows that

$$\begin{pmatrix} -0.671120 & 0.539127 & -0.604053 \\ -0.314952 & 0.647058 & 0.723638 \\ 0.671120 & 0.539127 & -0.264454 \end{pmatrix} = (F \ J \ F) \begin{pmatrix} -0.762440 & 0.474554 & -0.547197 \\ -0.357808 & 0.569558 & 0.767534 \\ 0.539127 & 0.671120 & -0.264454 \end{pmatrix}$$

which on solving gives

$$(F \ J \ F) = \begin{pmatrix} 0.990534 & -0.035103 & 0.132700 \\ 0.021102 & 0.994197 & 0.105482 \\ -0.135633 & -0.101683 & 0.985527 \end{pmatrix}$$

which is a rotation of 9.89^0 about $[0.602879 \ -0.780887 \ 0.163563]_F$
The invariant-line strain $(F \ S \ F)=(F \ J \ F)(F \ B \ F)$ is thus

$$(F \ S \ F) = \begin{pmatrix} 1.125317 & -0.039880 & 0.106601 \\ 0.023973 & 1.129478 & 0.084736 \\ -0.154089 & -0.115519 & 0.791698 \end{pmatrix}$$

and we note that $(F \ S \ F)^{-1}=(F \ B \ F)^{-1}(F \ J \ F)^{-1}$ is given by

$$(F \ S \ F)^{-1} = \begin{pmatrix} 0.871896 & 0.018574 & -0.119388 \\ -0.030899 & 0.875120 & -0.089504 \\ 0.165189 & 0.131307 & 1.226811 \end{pmatrix}$$

Stage 2: Determination of the Orientation Relationship

The orientation relationship between the austenite and martensite is best expressed in terms of a co-ordinate transformation matrix $(\alpha \ J \ \gamma)$. Any vector $\underline{u}$ or any plane normal $\underline{h}$ can then be expressed in either crystal basis by using the equations

$$[\alpha;\underline{u}] = (\alpha \ J \ \gamma)[\gamma;\underline{u}]$$
$$[\gamma;\underline{u}] = (\gamma \ J \ \alpha)[\alpha;\underline{u}]$$
$$(\underline{h};\alpha^*) = (\underline{h};\gamma^*)(\gamma \ J \ \alpha)$$
$$(\underline{h};\gamma^*) = (\underline{h};\alpha^*)(\alpha \ J \ \gamma)$$

63

<u>Example 19: The Martensite-Austenite Orientation Relationship</u>

For the martensite reaction considered in example 18, determine the orientation relationship between the parent and product lattices.

The orientation relationship can be expressed in terms of a co-ordinate transformation matrix $(\alpha\ J\ \gamma)$, which is related to the transformation strain $(\gamma\ S\ \gamma)$ via eq.15h, so that

$$(\alpha\ J\ \gamma)(\gamma\ S\ \gamma) = (\alpha\ C\ \gamma)$$

where $(\alpha\ C\ \gamma)$ is the Bain correspondence matrix (eq.20a), and it follows that

$$(\alpha\ J\ \gamma) = (\alpha\ C\ \gamma)(\gamma\ S\ \gamma)^{-1}$$

In example 18 the matrices representing the transformation strain and its inverse were determined in the basis F, and can be converted into the basis γ by means of a similarity transformation. However, because $\underline{f}_i$ are parallel to $\underline{a}_i$, it can easily be demonstrated that $(F\ S\ F)=(\gamma\ S\ \gamma)$ and $(F\ S\ F)^{-1}=(\gamma\ S\ \gamma)^{-1}$. Hence, using the data from example 18 and the Bain correspondence matrix from eq.20a, we see that

$$(\alpha\ J\ \gamma) = \begin{pmatrix} 1 & \bar{1} & 0 \\ 1 & 1 & 0 \\ 0 & 0 & 1 \end{pmatrix} \begin{pmatrix} 0.871896 & 0.018574 & -0.119388 \\ -0.030899 & 0.875120 & -0.089504 \\ 0.165189 & 0.131307 & 1.226811 \end{pmatrix}$$

$$(\alpha\ J\ \gamma) = \begin{pmatrix} 0.902795 & -0.856546 & -0.029884 \\ 0.840997 & 0.893694 & -0.208892 \\ 0.165189 & 0.131307 & 1.226811 \end{pmatrix}$$

$$(\gamma\ J\ \alpha) = \begin{pmatrix} 0.582598 & 0.542718 & 0.106602 \\ -0.552752 & 0.576725 & 0.084736 \\ -0.019285 & -0.134804 & 0.791698 \end{pmatrix}$$

These co-ordinate transformation matrices can be used to show that

$$(1\ 1\ 1)_\gamma = (0.010561\ \ 0.984639\ \ 0.983036)_\alpha$$
and
$$[\bar{1}\ 0\ 1]_\gamma = [-0.932679\ \ -1.049889\ \ 1.061622]_\alpha$$

This means that $(1\ 1\ 1)_\gamma$ is very nearly parallel to $(0\ 1\ 1)_\alpha$ and $[\bar{1}\ 0\ 1]_\gamma$ is about 3° from $[\bar{1}\ \bar{1}\ 1]_\alpha$. The orientation relationship is illustrated in Fig. 22.

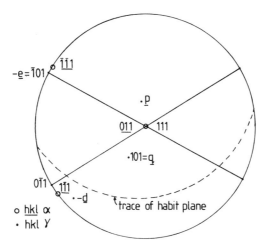

<u>Fig. 22</u>: Stereographic representation of the orientation relationship between martensite and austenite, as deduced in example 19. The lattice-invariant shear plane (<u>q</u>) and direction (-<u>e</u>), and the habit plane (<u>p</u>) and unit displacement vector (<u>d</u>) are also illustrated.

Stage 3: The Nature of the Shape Deformation

Having determined the transformation strain (F S F), it remains to factorise it according to eq.22a, (F S F)=(F P F)(F Q F), where (F P F) is the shape deformation matrix. This factorisation is not unique; (F P F) is physically significant because it can be experimentally determined and describes the macroscopic change in shape of the parent crystal, but the above equation also implies that (F Q F) operates before (F P F). This implication is not real since the transformation does not occur in two stages with (F Q F) followed by (F P F). All the changes necessary for transformation occur simultaneously at the moving interface. Transformation dislocations (atomic height steps) in the interface cause the FCC lattice to change to the BCC lattice as the interface moves and the deformation that this produces is described by (F S F). The intrinsic dislocations which lie along the invariant-line in the interface have Burgers vectors which are perfect lattice vectors of the parent lattice. They cannot therefore take part in the actual transformation of the lattice, but as the interface moves, they inhomogeneously shear the volume of the material swept by the interface [27]. This is of course the lattice-invariant shear discussed above, which in combination with the shape deformation due to (F S F) gives the experimentally observed IPS surface relief (F P F). This then is the physical interpretation of the transformation process.

The crystallographic theory of martensite is on the other hand called phenomenological; the steps into which the transformation is factorised (e.g., the two "shears" (F P F) & (F Q F)) are not unique and do not necessarily describe the actual path by which the atoms move from one lattice to the other. The theory simply provides a definite link between the initial and final states without being certain of the path in between.

Example 20: The Habit Plane and the Shape Deformation

For the martensite reaction considered in example 18, determine the habit plane of the martensite plate, assuming that the lattice-invariant shear occurs on the system $(1\ 0\ 1)_\gamma[\bar{1}\ 0\ 1]_\gamma$. Comment on the choice of this shear system and determine the nature of the shape deformation.

The lattice invariant shear is on $(1\ 0\ 1)[\bar{1}\ 0\ 1]$ and since its effect is to cancel the shape change due to (F Q F), the latter must be a shear on $(1\ 0\ 1)[1\ 0\ \bar{1}]$. To solve for the habit plane (unit normal $\underline{p}$) it is necessary to factorise (F S F) into the two invariant plane strains (F P F)=I + m[F;$\underline{d}$]($\underline{p}$;F*) and (F Q F)=I + n[F;$\underline{e}$]($\underline{q}$;F*).

The transformation strain (F S F) of example 18 was calculated by phenomenologically combining the Bain Strain with a rigid body rotation, with the latter chosen to make (F S F) an invariant line strain, subject to the condition that the invariant line $\underline{u}$ of (F S F) must lie in $(1\ 0\ 1)_\gamma$ and that the invariant-normal $\underline{h}$ of (F S F) must define a plane containing $[1\ 0\ \bar{1}]_\gamma$. This is of course compatible with the lattice-invariant shear system chosen in the present example since $\underline{u}.\underline{q}=\underline{e}.\underline{h}=0$. From eq.22a,

$$(F\ S\ F)=(F\ P\ F)(F\ Q\ F)=\{I + m[F;\underline{d}](\underline{p};F^*)\}\{I + n[F;\underline{e}](\underline{q};F^*)\} \qquad .....(24a)$$

and using eq.13, we see that

$$(F\ S\ F)^{-1} = (F\ Q\ F)^{-1}(F\ P\ F)^{-1}$$
$$=\{I - n[F;\underline{e}](\underline{q};F^*)\}\{I - am[F;\underline{d}](\underline{p};F^*)\} \qquad .....(24b)$$

where $(1/a)$=det(F P F) and det(F Q F)=1.

Using eq.24b, we obtain

$$(\underline{q};F^*)(F\ S\ F)^{-1} = (\underline{q};F^*) - b(\underline{p};F^*)$$

where b is a scalar constant given by b=am($\underline{q}$;F*)[F;$\underline{d}$].

Hence,
$$b(\underline{p};F^*) = (\underline{q};F^*) - (\underline{q};F^*)(F\ S\ F)^{-1}$$

$$= (0.707\ \ 0\ \ 0.707) - (0.707\ \ 0\ \ 0.707) \begin{pmatrix} 0.871896 & 0.018574 & -0.119388 \\ 0.030894 & 0.875120 & -0.089504 \\ 0.165189 & 0.131307 & 1.226811 \end{pmatrix}$$

$$= (-0.026223\ \ -0.105982\ \ -0.075960)$$

This can be normalised to give $\underline{p}$ as a unit vector:

$$(\underline{p};F^*)\|(0.197162\ \ 0.796841\ \ 0.571115)$$

($\underline{p}$;F*) of course represents the indices of the habit plane of the martensite plate. As expected, the habit plane is irrational. To completely determine the shape deformation matrix (F P F) we also need to know m and $\underline{d}$. Using eq.24a, we see that

$$(F\ S\ F)[F;\underline{e}] = [F;\underline{e}] + m[F;\underline{d}](\underline{p};F^*)[F;\underline{e}]$$

66

writing c as the scalar constant c=(p;F*)[F;e], we get

$$cm[F;\underline{d}] = (F\ S\ F)[F;\underline{e}] - [F;\underline{e}]$$

$$
= \begin{pmatrix} 1.125317 & -0.039880 & 0.106601 \\ 0.023973 & 1.129478 & 0.084736 \\ -0.154089 & -0.115519 & 0.791698 \end{pmatrix} \begin{pmatrix} 0.707107 \\ 0.000000 \\ -0.707107 \end{pmatrix} - \begin{pmatrix} 0.707107 \\ 0.000000 \\ -0.707107 \end{pmatrix}
$$

so that $\qquad cm[F;\underline{d}] = [0.013234\ -0.042966\ 0.038334]$

Now, c = (0.197162 0.796841 0.571115)[0.707107 0 -0.707107] = -0.26442478
so that m[F;$\underline{d}$] = [-0.050041 0.162489 -0.144971]
Since $\underline{d}$ is a unit vector, it can be obtained by normalising m$\underline{d}$ to give

$$[F;\underline{d}] = [-0.223961\ 0.727229\ -0.648829]$$
and $\qquad\qquad\qquad m=|m\underline{d}|= 0.223435$

The magnitude 'm' of the displacements involved can be factorised into a shear component 's' parallel to the habit plane and a dilatational component 'δ' normal to the habit plane. Hence, δ=m$\underline{d}$.p=0.0368161 and s=(m^2 - δ^2)$^{1/2}$=0.220381. These are typical values of the dilatational and shear components of the shape strain found in ferrous martensites.

The shape deformation matrix, using the above data, is given by

$$
(F\ P\ F)\ =\ \begin{pmatrix} 0.990134 & -0.039875 & -0.028579 \\ 0.032037 & 1.129478 & 0.092800 \\ -0.028583 & -0.115519 & 0.917205 \end{pmatrix}
$$

Stage 4: The Nature of the Lattice-Invariant Shear

We have already seen that the shape deformation (F P F) cannot account for the overall conversion of the FCC lattice to that of BCC martensite. An additional homogeneous lattice varying shear (F Q F) is necessary, which in combination with (F P F) completes the required change in structure. However, the macroscopic shape change observed experimentally is only due to (F P F); the effect of (F Q F) on the macroscopic shape change must thus be offset by a system of inhomogeneously applied lattice-invariant shears. Clearly, to macroscopically cancel the shape change due to (F Q F), the lattice-invariant shear must be the inverse of (F Q F). Hence, the lattice invariant shear operates on the plane with unit normal q but in the direction -e, its magnitude on average being the same as that of (F Q F).

Example 21: The Lattice Invariant Shear

For the martensite reaction discussed in examples 18-20, determine the nature of the homogeneous shear (F Q F) and hence deduce the magnitude of the lattice-invariant shear. Assuming that the lattice-invariant shear is a slip deformation, determine the spacing of the intrinsic dislocations in the habit plane, which are responsible for this inhomogeneous deformation.

Since (F S F)=(F P F)(F Q F), it follows that (F Q F)=(F P F)$^{-1}$(F S F), so that

$$\text{(F Q F)} = \begin{pmatrix} 1.009516 & 0.038459 & 0.027564 \\ -0.030899 & 0.875120 & -0.089505 \\ 0.027568 & 0.111417 & 1.079855 \end{pmatrix} \begin{pmatrix} 1.125317 & -0.039880 & 0.106601 \\ 0.023973 & 1.129478 & 0.084736 \\ -0.154089 & -0.115519 & 0.791698 \end{pmatrix}$$

$$\text{(F Q F)} = \begin{pmatrix} 1.132700 & 0.000000 & 0.132700 \\ 0.000000 & 1.000000 & 0.000000 \\ -0.132700 & 0.000000 & 0.867299 \end{pmatrix}$$

Comparison with eq.11d shows that this is a homogeneous shear on the system (1 0 1)[1 0 $\bar{1}$]$_F$ with a magnitude n=0.2654.

The lattice-invariant shear is thus determined since it is the inverse of (F Q F), having the same average magnitude but occurring inhomogeneously on the system (1 0 1)[$\bar{1}$ 0 1]$_F$. If the intrinsic interface dislocations which cause this shear have a Burgers vector $\underline{b}$=(a$_\gamma$/2)[$\bar{1}$ 0 1]$_\gamma$ and if they occur on every K'th slip plane, then if the spacing of the (1 0 1)$_\gamma$ planes is given by d, it follows that

$$n = |\underline{b}|/Kd = 1/K$$

so that

$$K = 1/0.2654 = 3.7679$$

Of course, K must be an integral number, and the non-integral result must be taken to mean that there will on average be a dislocation located on every 3.7679th slip plane; in reality, the dislocations will be non-uniformly placed, either 3 or 4 (1 0 1) planes apart.

The line vector of the dislocations is the invariant-line $\underline{u}$ and the spacing of the intrinsic dislocations, as measured on the habit plane is Kd/($\underline{u} \wedge \underline{p} . \underline{q}$) where all the vectors are unit vectors. Hence, the average spacing would be

$$3.7679(a_\gamma 2^{-1/2})/0.8395675 = 3.1734a_\gamma$$

and if a$_\gamma$=3.56A^0, then the spacing is 11.3A^0 on average.

If on the other hand, the lattice-invariant shear is a twinning deformation (rather than slip), then the martensite plate will contain very finely spaced transformation twins, the structure of the interface being radically different from that deduced above, since it will no longer contain any intrinsic dislocations. The mismatch between the parent and product lattices was in the slip case accommodated with the help of intrinsic dislocations, whereas for the internally twinned martensite there are no such dislocations. Each twin terminates in the interface to give a facet between the parent and product lattices, a facet which is forced into coherency. The width of the twin and the size of the facet is sufficiently small to enable this forced coherency to exist. The alternating twin related regions thus prevent misfit from accumulating over large distances along the habit plane.

If the (fixed) magnitude of the twinning shear is denoted 'S', then the volume fraction V of the twin orientation, necessary to cancel the effect of (F Q F), is given by V=n/S, assuming that n<S. In the above example, the lattice-invariant shear occurs on (1 0 1)[$\bar{1}$ 0 1]$_\gamma$ which corresponds to (1 1 2)$_\alpha$[$\bar{1}$ $\bar{1}$ 1]$_\alpha$, and twinning on this latter system involves a shear S=0.707107, giving V=0.2654/0.707107=0.375.

It is important to note that the twin plane in the martensite corresponds to a mirror plane in the austenite; this is a necessary condition when the lattice-invariant shear involves twinning. The condition arises because the twinned and untwinned regions of the martensite must undergo Bain Strain along different though crystallographically equivalent principal axes [2,4].

68

The above theory clearly predicts a certain volume fraction of twins in each martensite plate, when the lattice-invariant shear is twinning as opposed to slip. However, the factors governing the spacing of the twins are less quantitatively established; the finer the spacing of the twins, the lower will be the strain energy associated with the matching of each twin variant with the parent lattice at the interface. On the other hand, the amount of coherent twin boundary within the martensite increases as the spacing of the twins decreases.

A factor to bear in mind is that the lattice-invariant shear is an integral part of the transformation; it does not happen as a separate event after the lattice change has occurred. The transformation and the lattice-invariant shear all occur simultaneously at the interface, as the latter migrates. It is well known that in ordinary plastic deformation, twinning rather than slip tends to be the favoured deformation mode at low temperatures or when high strain rates are involved. It is therefore often suggested that martensite with low M_s temperatures will tend to be twinned rather than slipped, but this cannot be formally justified because the lattice-invariant shear is an integral part of the transformation and not a physical deformation mode on its own. Indeed, it is possible to find lattice-invariant deformation modes in martensite which do not occur in ordinary plastic deformation experiments. The reasons why some martensites are internally twinned and others slipped are not clearly understood [49]. When the spacing of the transformation twins is roughly comparable to that of the dislocations in slipped martensite, the interface energies are roughly equal. The interface energy increases with twin thickness and at the observed thicknesses is very large compared with the corresponding interface in slipped martensite. The combination of the relatively large interface energy and the twin boundaries left in the martensite plate means that internally twinned martensite is never thermodynamically favoured relative to slipped martensite. It is possible that kinetic factors such as interface mobility actually determine the type of martensite that occurs.

INTERFACES IN CRYSTALLINE SOLIDS

Atoms in the boundary between crystals must in general be displaced from positions they would occupy in the undisturbed crystal, but it is now well established that many interfaces have a periodic structure. In such cases, the misfit between the crystals connected by the boundary is not distributed uniformly over every element of the interface; it is periodically localised into discontinuities which separate patches of the boundary where the fit between the two crystals is good or perfect. When these discontinuities are well separated, they may individually be recognised as interface dislocations which separate coherent patches in the boundary, which is macroscopically said to be semi-coherent. Stress-free coherent interfaces can of course only exist between crystals which can be related by a transformation strain which is an invariant-plane strain. This transformation strain may be real or notional as far as the calculation of the interface structure is concerned, but a real strain implies the existence of an atomic correspondence (and an associated macroscopic shape change of the transformed region) between the two crystals, which a notional strain does not.

Incoherency presumably sets in when the misfit between adjacent crystals is so high that it cannot satisfactorily be localised into identifiable interface dislocations, giving a boundary structure which is difficult to physically interpret, other than to say that the motion of such an interface must always occur by the unco-ordinated and haphazard transfer of atoms across the interface. This could be regarded as a definition of incoherency; as will become clear later, the intuitive feeling that all "high-angle" boundaries are incoherent is not correct.

The misfit across an interface can formally be described in terms of the net Burgers vector $\underline{b}_t$ crossing a vector $\underline{p}$ in the interface[50,51,5]. If this misfit is sufficiently small then the boundary structure may relax into a set of discrete interfacial dislocations (where the misfit is concentrated) which are separated by patches of good fit.

In any case, $\underline{b}_t$ may be deduced by constructing a Burgers circuit across the interface, and examining the closure failure when a corresponding circuit is constructed in a perfect reference lattice. The procedure is illustrated in Fig. 23, where crystal A is taken to be the reference lattice. An initial right-handed Burgers circuit OAPBO is constructed such that it straddles the interface across any vector $\underline{p}$ in the interface ($\underline{p}$ = OP); the corresponding circuit in the perfect reference lattice is constructed by deforming the crystal B (of the bi-crystal A-B) in such a way that it is converted into the lattice of A, eliminating the interface. If the deformation (A S A) converts the reference lattice into the B lattice, then the inverse deformation (A S A)$^{-1}$ converts the bi-crystal into a single A crystal, and the Burgers circuit in the perfect reference lattice becomes OAPP'B, with a closure failure PP', which is of course identified as $\underline{b}_t$. Inspection of the vectors forming the triangle OPP' of Fig. 23b shows that:

$$[A;\underline{b}_t] = \{ \, I - (A \ S \ A)^{-1} \} \, [A;\underline{p}] \qquad \qquad .....(25)$$

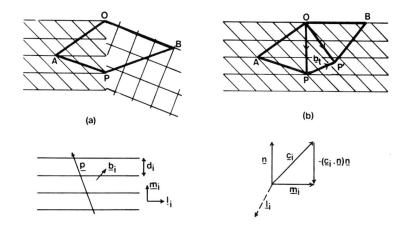

Fig. 23: (a,b) Burgers circuit used to define the formal dislocation content of an interface[5], (c) the vector $\underline{p}$ in the interface, (d) relationship between $\underline{l}_i$, $\underline{m}_i$, $\underline{c}_i$ and $\underline{n}$.

Hence, the net Burgers vector content $\underline{b}_t$ crossing an arbitrary vector $\underline{p}$ in the interface is formally given by eq.25.

The misfit in any interface can in general be accommodated with three arrays of interfacial dislocations, whose Burgers vectors $\underline{b}_i$ (i = 1,2,3) form a non-coplanar set.

Hence, $\underline{b}_t$ can in general be factorised into three arrays of interfacial dislocations, each array with Burgers vector $\underline{b}_i$, unit line vector $\underline{l}_i$ and array spacing d_i , the latter being measured in the interface plane. If the unit interface normal is $\underline{n}$, then a vector $\underline{m}_i$ may be defined as (the treatment that follows is due to Knowles [52] and Read [53]):

$$\underline{m}_i = \underline{n} \wedge \underline{l}_i / d_i \qquad \qquad .....(26a)$$

We note the $|\underline{m}_i| = 1/d_i$, and that any vector $\underline{p}$ in the interface crosses $(\underline{m}_i.\underline{p})$ dislocations of type i (see Fig. 23c). Hence, for the three kinds of dislocations we have

$$\underline{b}_t = (\underline{m}_1.\underline{p})\underline{b}_1 + (\underline{m}_2.\underline{p})\underline{b}_2 + (\underline{m}_3.\underline{p})\underline{b}_3$$

We note that the Burgers vectors $(\underline{b}_i)$ of interface dislocations are generally lattice translation vectors of the reference lattice in which they are defined. This makes them "perfect" in the sense that the displacement of one of the crystals through $\underline{b}_i$ relative to the other does not change the structure of the boundary. On the basis of elastic strain energy arguments, $\underline{b}_i$ should be as small as possible. On substituting this into eq.25, we get:

$$(\underline{m}_1;A^*)[A;\underline{p}][A;\underline{b}_1] + (\underline{m}_2;A^*)[A;\underline{p}][A;\underline{b}_2] + (\underline{m}_3;A^*)[A;\underline{p}][A;\underline{b}_3] = (A \ T \ A)[A;\underline{p}] \qquad .....(26b)$$

where $(A \ T \ A) = I - (A \ S \ A)^{-1}$.

71

If a scalar dot product is taken on both sides of eq.26b with the vector $\underline{b}_1^*$, where $\underline{b}_1^*$ is

$$\underline{b}_1^* = \underline{b}_2 \wedge \underline{b}_3 \ / \ (\underline{b}_1 . \underline{b}_2 \wedge \underline{b}_3) \qquad .....(26c)$$

then we obtain

$$(\underline{b}_1^*;A^*)(A \ T \ A)[A;\underline{p}] = (\underline{m}_1;A^*)[A;\underline{p}] \qquad .....(26d)$$

If $\underline{p}$ is now taken to be equal to $\underline{l}_1$ in eq.26d, we find that

$$(\underline{b}_1^*;A^*)(A \ T \ A)[A;\underline{l}_1] = 0$$

or

$$(\underline{l}_1;A)(A \ T' \ A)[A^*;\underline{b}_1^*] = 0 \qquad .....(26e)$$

If we define a new vector $\underline{c}_1$ such that

$$[A^*;\underline{c}_1] = (A \ T' \ A)[A^*;\underline{b}_1^*] \qquad .....(26f)$$

then eq.26e indicates that $\underline{c}_1$ is normal to $\underline{l}_1$. Furthermore, if $\underline{m}_1$ is now substituted for $\underline{p}$ in eq.26d, then we find:

$$(\underline{b}_1^*;A^*)(A \ T \ A)[A;\underline{m}_1] = (\underline{m}_1;A^*)[A;\underline{m}_1] = |\underline{m}_1|^2$$

or
$$(\underline{m}_1;A)(A \ T' \ A)[A^*;\underline{b}_1^*] = |\underline{m}_1|^2$$

or
$$(\underline{m}_1;A)[A^*;\underline{c}_1] = |\underline{m}_1|^2$$

or
$$\underline{m}_1 . \underline{c}_1 = |\underline{m}_1|^2$$

These equations indicate that the projection of $\underline{c}_1$ along $\underline{m}_1$ is equal to the magnitude of $\underline{m}_1$. Armed with this and the earlier result that $\underline{c}_1$ is normal to $\underline{l}_1$, we may construct the vector diagram illustrated in Fig. 23d, which illustrates the relations between $\underline{m}_1$, $\underline{l}_1$, $\underline{n}$ and $\underline{c}_1$. From this diagram, we see that

$$\underline{m}_1 = \underline{c}_1 - (\underline{c}_1 . \underline{n})\underline{n} \qquad .....(27a)$$

so that
$$|\underline{m}_1 \wedge \underline{n}| = |\underline{c}_1 \wedge \underline{n}| = 1/d_1 \qquad .....(27b)$$

and in addition, Fig. 23d shows that

$$\underline{l}_1 \ \| \ \underline{c}_1 \wedge \underline{n} \qquad .....(27c)$$

alternatively,
$$(1/d_1)\underline{l}_1 = \underline{c}_1 \wedge \underline{n} \qquad .....(27d)$$

The relations of the type developed in eq.27 are, after appropriately changing indices, also applicable to the other two arrays of interfacial dislocations, so that we have achieved a way of deducing the line vectors and array spacings to be found in an interface (of unit normal $\underline{n}$) connecting two arbitrary crystals A and B, related by the deformation (A S A) which transforms the crystal A to B.

Example 22: The Symmetrical Tilt Boundary

Given that the Burgers vectors of interface dislocations in low-angle boundaries are of the form <1 0 0>, calculate the dislocation structure of a symmetrical tilt boundary formed between two grains, A and B, related by a rotation of 2θ about the [1 0 0] axis. The crystal structure of the grains is simple cubic, with a lattice parameter of 1 angstrom.

The definition of a tilt boundary is that the unit boundary normal $\underline{n}$ is perpendicular to the axis of rotation which generates one crystal from the other; a symmetrical tilt boundary has the additional property that the lattice of one crystal can be generated from the other by reflection across the boundary plane.

If we choose the orthonormal bases A and B (with basis vectors parallel to the cubic unit cell edges) to represent crystals A and B respectively, and also arbitrarily choose B to be the reference crystal, then the deformation which generates the A crystal from B is a rigid body rotation (B J B) consisting of a rotation of 2θ about $[1\ 0\ 0]_B$. Hence, (B J B) is given by (see eq.8c):

$$(B\ J\ B)\ =\ \begin{pmatrix} 1 & 0 & 0 \\ 0 & \cos(2\theta) & -\sin(2\theta) \\ 0 & \sin(2\theta) & \cos(2\theta) \end{pmatrix}$$

and $(B\ T\ B) = I - (B\ J\ B)^{-1}$ is given by

$$(B\ T\ B)\ =\ \begin{pmatrix} 0 & 0 & 0 \\ 0 & 1-\cos(2\theta) & \sin(2\theta) \\ 0 & -\sin(2\theta) & 1-\cos(2\theta) \end{pmatrix}$$

and

$$(B\ T'\ B)\ =\ \begin{pmatrix} 0 & 0 & 0 \\ 0 & 1-\cos(2\theta) & -\sin(2\theta) \\ 0 & \sin(2\theta) & 1-\cos(2\theta) \end{pmatrix}$$

Taking $[B;\underline{b}_1] = [1\ 0\ 0]$, $[B;\underline{b}_2] = [0\ 1\ 0]$ and $[B;\underline{b}_3] = [0\ 0\ 1]$, from eq.26c we note that $[B^*;\underline{b}_1^*] = [1\ 0\ 0]$, $[B^*;\underline{b}_2^*] = [0\ 1\ 0]$ and $[B^*;\underline{b}_3^*] = [0\ 0\ 1]$. Eq.26f can now be used to obtain the vectors $\underline{c}_i$:

$[B^*;\underline{c}_1] = (B\ T'\ B)[B^*;\underline{b}_1^*] = [0\ 0\ 0]$

$[B^*;\underline{c}_2] = (B\ T'\ B)[B^*;\underline{b}_2^*] = [0 \quad 1-\cos(2\theta) \quad \sin(2\theta)]$

$[B^*;\underline{c}_3] = (B\ T'\ B)[B^*;\underline{b}_3^*] = [0 \quad -\sin(2\theta) \quad 1-\cos(2\theta)]$

Since $\underline{c}_1$ is a null vector, dislocations with Burgers vector $\underline{b}_1$ do not exist in the interface; furthermore, since $\underline{c}_1$ is always a null vector, irrespective of the boundary orientation $\underline{n}$, this conclusion remains valid for any $\underline{n}$. This situation arises because $\underline{b}_1$ happens to be parallel to the rotation axis, and because (B J B) is an invariant-line strain, the invariant line being the rotation axis; since any two crystals of identical structure can always be related by a transformation which is a rigid body rotation, it follows that all grain boundaries (as opposed to phase boundaries) *need* only contain two sets of interface dislocations. If $\underline{b}_1$ had not been parallel to the rotation axis, then $\underline{c}_1$ would be finite and three sets of dislocations would be necessary to accommodate the misfit in the boundary.

To calculate the array spacings d_i it is necessary to express $\underline{n}$ in the B^* basis. A symmetrical tilt boundary always contains the axis of rotation and has the same indices in both bases. It follows that

$$(\underline{n};B^*) = (0 \quad \cos(\theta) \quad -\sin(\theta))$$

and from eq.27d,

$$(1/d_2)\underline{l}_2 \ = \underline{c}_2 \wedge \underline{n} = [-2\sin(\theta) \quad 0 \quad 0]_B$$

so that

$$[B;\underline{l}_2] = [-1\ 0\ 0]$$

and

$$d_2 \ = \ 1/2\sin(\theta).$$

Similarly,
$$(1/d_3)\underline{l}_3 = \underline{c}_3 \wedge \underline{n} = [0\ 0\ 0]$$
so that dislocations with Burgers vector $\underline{b}_3$ have an infinite spacing in the interface, which therefore consists of just one set of interface dislocations with Burgers vector $\underline{b}_2$.

These results are of course identical to those obtained from a simple geometrical construction of the symmetrical tilt boundary[5]. We note that the boundary is glissile (i.e., its motion does not require the creation or destruction of lattice sites) because $\underline{b}_2$ lies outside the boundary plane, so that the dislocations can glide conservatively as the interface moves. In the absence of diffusion, the movement of the boundary leads to a change in shape of the "transformed" region, a shape change described by (B J B) when the boundary motion is towards the crystal A.

If on the other hand, the interface departs from its symmetrical orientation (without changing the orientation relationship between the two grains), then the boundary ceases to be glissile, since the dislocations with Burgers vector $\underline{b}_3$ acquire finite spacings in the interface. Such a boundary is called an asymmetrical tilt boundary. For example, if $\underline{n}$ is taken to be $(\underline{n};B^*) = (0\ 1\ 0)$, then

$$[B;\underline{l}_2] = [\bar{1}\ 0\ 0] \text{ and } d_2 = 1/\sin(2\theta)$$
$$[B;\underline{l}_3] = [\bar{1}\ 0\ 0] \text{ and } d_3 = 1/(1-\cos(2\theta))$$

The edge dislocations with Burgers vector $\underline{b}_3$ lie in the interface plane and therefore have to climb as the interface moves. This renders the interface sessile.

Finally, we consider the structure of a twist boundary, a boundary where the axis of rotation is parallel to $\underline{n}$. Taking $(\underline{n};B^*) = (1\ 0\ 0)$, we find:

$$[B;\underline{l}_2] \parallel [0 \quad \sin(2\theta) \quad \cos(2\theta)-1], \text{ and } d_2 = [2-2\cos(2\theta)]^{-0.5}$$

and

$$[B;\underline{l}_3] \parallel [0 \quad 1-\cos(2\theta) \quad \sin(2\theta)], \text{ and } d3 = d2$$

We note that dislocations with Burgers vector $\underline{b}_1$ do not exist in the interface since $\underline{b}_1$ is parallel to the rotation axis. The interface thus consists of a cross grid of two arrays of dislocations with Burgers vectors $\underline{b}_2$ and $\underline{b}_3$ respectively, the array spacings being identical. Although it is usually stated that pure twist boundaries contain grids of pure screw dislocations, we see that both sets of dislocations actually have a small edge component. This is because the dislocations, where they mutually intersect, introduce jogs into each other so that the line vector in the region between the points of intersection does correspond to a pure screw orientation, but the jogs make the macroscopic line vector deviate from this screw orientation.

Example 23: The interface between alpha and beta brass

The lattice parameter of the FCC alpha phase of a 60 wt.pct. Cu-Zn alloy is 3.6925 angstroms, and that of the BCC beta phase is 2.944 angstroms [54]. Two adjacent grains A and B (alpha phase and beta phase respectively) are orientated in such a way that

$$[1\ 1\ 1]_A \parallel [1\ 1\ 0]_B$$
$$[1\ 1\ \bar{2}]_A \parallel [\bar{1}\ 1\ 0]_B$$
$$[1\ \bar{1}\ 0]_A \parallel [0\ 0\ \bar{1}]_B$$

and the grains are joined by a boundary which is parallel to $(\bar{1}\ 1\ 0)_A$. Assuming that the misfit in this interface can be fully accommodated by interface dislocations which have Burgers vectors $[A;\underline{b}_1] = 0.5[1\ 0\ \bar{1}]$, $[A;\underline{b}_2] = 0.5[0\ 1\ \bar{1}]$ and $[A;\underline{b}_3] = 0.5[1\ 1\ 0]$, calculate the misfit dislocation structure of the interface. Also assume that the smallest pure deformation which relates FCC and BCC crystals is the Bain strain.

The orientation relations provided are first used to calculate the co-ordinate transformation matrix (B J A), using the procedure given in example 4 and 5. (B J A) is thus found to be:

$$(B \ J \ A) \ = \ \begin{pmatrix} 0.149974 & 0.149974 & 1.236185 \\ 0.874109 & 0.874109 & -0.212095 \\ -0.886886 & 0.886886 & 0.000000 \end{pmatrix}$$

The smallest pure deformation relating the two lattices is stated to be the Bain strain, but the total transformation (A S A), which carries the A lattice to that of B, may include an additional rigid body rotation. Determination of the interface structure requires a knowledge of (A S A), which may be calculated from the equation $(A \ S \ A)^{-1} = (A \ C \ B)(B \ J \ A)$, where (A C B) is the correspondence matrix. Since (B J A) is known, the problem reduces to the determination of the correspondence matrix; if a vector equal to a basis vector of B, due to transformation becomes a new vector $\underline{u}$, then the components of $\underline{u}$ in the basis A form one column of the correspondence matrix.

For the present example, the correspondence matrix must be a variant of the Bain correspondence, so that we know that $[1 \ 0 \ 0]_B$ must become, as a result of transformation, either a vector of the form $<1 \ 0 \ 0>_A$, or a vector of the form $(0.5)<1 \ 1 \ 0>_A$, although we do not know its final specific indices in the austenite lattice. However, from the matrix (B J A) we note that $[1 \ 0 \ 0]_B$ is close to $[0 \ 0 \ 1]_A$, so that it is reasonable to assume that $[1 \ 0 \ 0]_B$ corresponds to $[0 \ 0 \ 1]_A$, so that the first column of (A C B) is $[0 \ 0 \ 1]$. Similarly, using (B J A) we find that $[0 \ 1 \ 0]_B$ is close to $[1 \ 1 \ 0]_A$ and $[0 \ 0 \ 1]_B$ is close to $[\bar{1} \ 1 \ 0]_A$, so that the other two columns of (A C B) are $[0.5 \ 0.5 \ 0]$ and $[-0.5 \ 0.5 \ 0]$. Hence, (A C B) is found to be:

$$(A \ C \ B) \ = \ (1/2) \begin{pmatrix} 0 & 1 & -1 \\ 0 & 1 & 1 \\ 2 & 0 & 0 \end{pmatrix}$$

so that $(A \ S \ A)^{-1}=(A \ C \ B)(B \ J \ A)$ is given by:

$$(A \ S \ A)^{-1} \ = \ \begin{pmatrix} 0.880500 & -0.006386 & -0.106048 \\ -0.006386 & 0.880500 & -0.106048 \\ 0.149974 & 0.149974 & 1.236183 \end{pmatrix}$$

and since $(A \ T \ A) = I - (A \ S \ A)^{-1}$, (A T' A) is given by:

$$(A \ T' \ A) = \begin{pmatrix} 0.119500 & 0.006386 & -0.149974 \\ 0.006386 & 0.119500 & -0.149974 \\ 0.016048 & 0.016048 & -0.236183 \end{pmatrix}$$

The vectors $\underline{b}_i^*$, as defined by eq.26c, are given by:

$$[A^*;\underline{b}_1^*] = [1 \ \bar{1} \ \bar{1}]$$
$$[A^*;\underline{b}_2^*] = [\bar{1} \ 1 \ \bar{1}]$$
$$[A^*;\underline{b}_3^*] = [1 \ 1 \ 1]$$

75

and using eq.26f, we find

$$[A^*;\underline{c}_1] = [0.263088 \quad 0.036860 \quad 0.236183]$$
$$[A^*;\underline{c}_2] = [0.036860 \quad 0.263088 \quad 0.236083]$$
$$[A^*;\underline{c}_3] = [-0.024088 \quad -0.024088 \quad -0.024087]$$

and from eq.27d and the fact that $(\underline{n};A^*) = 3.6925(-0.707107 \quad 0.707107 \quad 0)$,

$$(1/d_1)[A;\underline{l}_1] = [-0.012249 \quad -0.012249 \quad 0.015556]$$
$$(1/d_2)[A;\underline{l}_2] = [-0.012249 \quad -0.012249 \quad 0.015556]$$
$$(1/d_3)[A;\underline{l}_3] = [\ 0.001249 \quad 0.001249 \quad -0.002498]$$

so that

$$d_1 = d_2 = 11.63A^o, \text{ and } d_3 = 88.52A^o$$

Coincidence Site Lattices

We have emphasized that eq.25 defines the formal (or mathematical) Burgers vector content of an interface, and it is sometimes possible to interpret this in terms of physically meaningful interface dislocations, if the misfit across the interface is sufficiently small. For high-angle boundaries, the predicted spacings of dislocations may turn out to be so small that the misfit is highly localised with respect to the boundary, and the dislocation model of the interface has only formal significance (it is often said that the dislocations get so close to each other that their cores overlap). The arrangement of atoms in such incoherent boundaries may be very haphazard, with little correlation of atomic positions across the boundary.

On the other hand, it is unreasonable to assume that all high-angle boundaries have the disordered structure suggested above. There is clear experimental evidence which shows that certain high-angle boundaries exhibit the characteristics of low-energy coherent or semi-coherent interfaces; for example, they exhibit strong facetting, have very low mobility in pure materials and the boundary diffusion coefficient may be abnormally low. These observations imply that at certain special relative crystal orientations, which would usually be classified as high-angle orientations, it is possible to obtain boundaries which have a distinct structure - they contain regions of good fit, which occur at regular intervals in the boundary plane, giving a pattern of good fit points in the interface. It is along these points that the two crystals connected by the boundary match exactly.

If we consider the good fit points to correspond to lattice points in the interface which are common to both crystals, then the following procedure allows us to deduce the pattern and frequency of these points for any given orientation relationship. If the two lattices (with a common origin) are notionally allowed to interpenetrate and fill all space, then there may exist lattice points (other than the origin) which are common to both the crystals. The set of these coincidence points forms a coincidence site lattice (CSL) [56,57], and the fractions of lattice points which are also coincidence sites is a rational fraction $1/\Sigma$. Σ is thus the reciprocal density of coincidence sites relative to ordinary lattice sites. The value of Σ is a function of the relative orientation of the two grains and not of the orientation of the boundary plane. The boundary simply intersects the CSL and will contain regions of good fit which have a periodicity corresponding to the periodicity of a planar net of the CSL along which the intersection occurs. Boundaries parallel to low-index planes of the CSL are two dimensionally periodic with relatively small repeat cells, and those boundaries with a high planar coincidence site density should have a relatively low energy.

Example 24: Coincidence site lattices

The axis-angle pair describing the orientation relationship between the two grains (A and B) of austenite is given by:

axis of rotation parallel to $[1\ 1\ 2]_A$

right-handed angle of rotation 180°

Show that a CSL with $\Sigma = 3$ can be formed by allowing the two lattices to notionally interpenetrate. Also show that 1 in 3 of the sites in an interface parallel to $(0\ \bar{2}\ 1)_A$ are coincidence sites.

Because of the centrosymmetric nature of the austenite lattice, a rotation of 180° about $[1\ 1\ 2]_A$, which generates the B grain from A, is equivalent to a reflection of the A lattice about $(1\ 1\ 2)_A$. We may therefore imagine that B is generated by reflection of the A lattice about $(1\ 1\ 2)_A$ as the mirror plane. The stacking sequence of $(1\ 1\ 2)_A$ is ....ABCDEFABCDEFABCDEF.... and grain A is represented below as a stack of $(1\ 1\ 2)$ planes, B being generated by reflecting the A lattice about one of the $(1\ 1\ 2)_A$ planes.

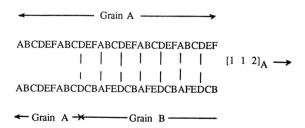

The first sequence represents a stack of $(1\ 1\ 2)_A$ planes, as do the first 9 layers of the second sequence. The remainder of the second sequence represents a stack of $(1\ 1\ 2)_B$ planes (note that B lattice is obtained by reflection of A about the 9th layer of the second sequence). Since the two sequences have the same origin, comparison of the first sequence with the B part of the second sequence amounts to allowing the two crystals to interpenetrate in order to identify coincidences. Clearly, every 3rd layer of grain B coincides exactly with a layer from the A lattice (dashed vertical lines), giving $\Sigma = 3$.

A boundary parallel to $(1\ 1\ 2)_A$ will be fully coherent; at least 1 in 3 of the sites in any other boundary, such as $(0\ \bar{2}\ 1)_A$ will be coincidence sites.

We now consider a mathematical method [5,58] of determining the CSL formed by allowing the lattices of crystals A and B to notionally interpenetrate; A and B are assumed to be related by a transformation (A S A) which deforms the A lattice into that of B; A and B need not have the same crystal structure or lattice parameters, so that (A S A) need not be a rigid body rotation.

Consider a vector $\underline{u}$ which is a lattice vector whose integral components do not have a common factor. As a result of the transformation (A S A), $\underline{u}$ becomes a new vector $\underline{w}$ such that

$$[A;\underline{w}] = (A\ S\ A)[A;\underline{u}] \qquad .....(28)$$

Of course, $\underline{w}$ does not necessarily have integral components in the A basis (i.e., it need not be a lattice vector of A). CSL vectors, on the other hand, identify lattice points which are common to both A and B, and therefore are lattice vectors of both crystals. It follows that CSL vectors have integral indices when referred to either crystal. Hence, $\underline{w}$ is only a CSL vector if it has integral components in the basis A. We note that $\underline{w}$ always has integral components in B, because a lattice vector of A (such as $\underline{u}$) always deforms into a lattice vector of B.

The meaning of Σ is that $1/\Sigma$ of the lattice sites of A or B are common to both A and B. It follows that any primitive lattice vector of A or B, when multiplied by Σ, must give a CSL vector. $\Sigma\underline{w}$ must therefore always be a CSL vector and if eq.28 is multiplied by Σ, then we obtain an equation in which the vector $\underline{u}$ always transforms into a CSL vector:

$$\Sigma[A;\underline{w}] = \Sigma(A\ S\ A)[A;\underline{u}] \qquad .....(29)$$

i.e., given that $\underline{u}$ is a lattice vector of A, whose components have no common factor, $\Sigma\underline{w}$ is a CSL vector with integral components in either basis. This can only be true if the matrix $\Sigma(A\ S\ A)$ has elements which are all integral since it is only then that $\Sigma[A;\underline{w}]$ has elements which are all integral.

It follows that if an integer H can be found such that all the elements of the matrix $H(A\ S\ A)$ are integers (without a common factor), then H is the Σ value relating A and B.

Applying this to the problem of example 24, the rotation matrix corresponding to the rotation 180^0 about $[1\ 1\ 2]_A$ is given by (eq.8c)

$$(A\ J\ A) = (1/3) \begin{pmatrix} \bar{2} & 1 & \bar{2} \\ 1 & \bar{2} & 2 \\ 2 & 2 & 1 \end{pmatrix}$$

and since 3 is the integer which when multiplied with $(A\ J\ A)$ gives a matrix of integral elements (without a common factor), the Σ value for this orientation is given by $\Sigma = 3$. For reasons of symmetry (see chapter 2), the above rotation is crystallographically equivalent to a rotation of 60^0 about $[1\ 1\ 1]_A$ and for this the rotation matrix is given by

$$(A\ J\ A) = (1/3) \begin{pmatrix} 2 & 2 & \bar{1} \\ \bar{1} & 2 & 2 \\ 2 & \bar{1} & 2 \end{pmatrix}$$

so that a rotation of 60^0 about $[1\ 1\ 1]_A$ also corresponds to a $\Sigma = 3$ value.

Finally, we see from eq.29 that if the integer H (defined such that $H(A\ S\ A)$ has integral elements with no common factor) turns out to be even, then the Σ value is obtained by successively dividing H by 2 until the result H' is an odd integer. H' then represents the true Σ value. This is because (eq.29) if $H[A;\underline{w}]$ is a CSL vector and if H is even then $H\underline{w}$ has integral even components in A, but $H'\underline{w}$ would also have integral components in A and would therefore represent a smaller CSL vector. From example 16, the transformation strain relating FCC-austenite and HCP-martensite is given by:

$$(Z\ P\ Z) = \begin{pmatrix} 1.083333 & 0.083333 & 0.083333 \\ 0.083333 & 1.083333 & 0.083333 \\ -0.166667 & -0.166667 & 0.833333 \end{pmatrix} = (1/12) \begin{pmatrix} 13 & 1 & 1 \\ 1 & 13 & 1 \\ -2 & -2 & 9 \end{pmatrix}$$

so that $H = 12$, but $H' = \Sigma = 3$. We can further illustrate this result by following the procedure

78

of example 24. The first sequence below represents a stack of the (1 1 1) planes of the FCC lattice, as do the first 9 planes of the second stacking sequence. The other planes of the second sequence represent the basal planes of the HCP lattice. Since two out of every 6 layers are in exact coincidence, $\Sigma = 3$, as shown earlier.

$$\longleftarrow \text{FCC austenite} \longrightarrow$$

ABCABCABCABCABCABCABCABCABC

ABCABCABCACACACACACACACACAC $\quad [1\ 1\ 1]_\gamma \| [0\ 0\ 0\ 1]_{HCP}$

$$\longleftarrow \text{FCC} \longrightarrow\hspace{-0.3em}\times\hspace{-0.5em}\longrightarrow \text{HCP} \longrightarrow$$

<u>Example 25: Symmetry and the Axis-Angle representations of CSL's</u>

Show that the coincidence site lattice associated with two cubic crystals related by a rotation of 50.5^o about <1 1 0>has $\Sigma = 11$. Using the symmetry operations of the cubic lattice, generate all possible axis-angle pair representations which correspond to this Σ value.

The rotation matrix corresponding to the orientation relation 50.5^o about <1 1 0> is given by (eq.8c):

$$(A\ J\ A) = \begin{pmatrix} 0.545621 & -0.545621 & 0.636079 \\ 0.181961 & 0.818039 & 0.545621 \\ 0.818039 & 0.181961 & -0.545621 \end{pmatrix} = (1/11)\begin{pmatrix} 6 & -6 & 7 \\ 2 & 9 & 6 \\ 9 & 2 & -6 \end{pmatrix}$$

so that $\Sigma = 11$ (A is the basis symbol representing one of the cubic crystals). The 24 symmetry operations of the cubic lattice are given by:

Angle (degrees)	Axis
0,90,180,270	<1 0 0>
90,180,270	<0 1 0>
90,180,270	<0 0 1>
180	<1 1 0>
180	<1 0 1>
180	<0 1 1>
180	<1 $\bar{1}$ 0>
180	<1 0 $\bar{1}$>
180	<0 1 $\bar{1}$>
120,240	<1 1 1>
120,240	<$\bar{1}$ $\bar{1}$ 1>
120,240	<1 $\bar{1}$ $\bar{1}$>
120,240	<$\bar{1}$ 1 $\bar{1}$>

If any symmetry operation is represented as a rotation matrix which then premultiplies (A J A), then the resulting new rotation matrix gives another axis-angle pair representation. Hence, the alternative axis-angle pair representations of $\Sigma = 11$ are found to be:

Angle (degrees)	Axis
82.15	<1 3 3>
162.68	<3 3 5>
155.37	<1 2 4>
180.00	<2 3 3>
62.96	<1 1 2>
129.54	<1 1 4>
129.52	<0 1 1>
126.53	<1 3 5>
180.00	<1 1 3>
100.48	<0 2 3>
144.89	<0 1 3>

The above procedure can be used to derive all the axis-angle pair representations of any Σ value, and the table below gives some of the CSL relations for cubic crystals, quoting the axis-angle pair representations which have the minimum angle of rotation, and also those corresponding to twin axes.

Table 3: Some CSL relations for Cubic crystals[5]

Σ	Angle	Axis	Twin axes
3	60.0	<1 1 1>	<1 1 1>, <1 1 2>
5	36.9	<1 0 0>	<0 1 2>, <0 1 3>
7	38.2	<1 1 1>	<1 2 3>
9	38.9	<1 1 0>	<1 2 2>, <1 1 4>
11	50.5	<1 1 0>	<1 1 3>, <2 3 3>
13a	22.6	<1 0 0>	<0 2 3>, <0 1 5>
13b	27.8	<1 1 1>	<1 3 4>
15	48.2	<2 1 0>	<1 2 5>
17a	28.1	<1 0 0>	<0 1 4>, <0 3 5>
17b	61.9	<2 2 1>	<2 2 3>, <3 3 4>
19a	26.5	<1 1 0>	<1 3 3>, <1 1 6>
19b	46.8	<1 1 1>	<2 3 5>
21a	21.8	<1 1 1>	<2 3 5>, <1 4 5>
21b	44.4	<2 1 1>	<1 2 4>

We note the following further points about CSL relations:

(i) All of the above CSL relations can be represented by a rotation of 180° about some rational axis which is not an even axis of symmetry. Any such operation corresponds to a twinning orientation (for centrosymmetric crystals), the lattices being reflected about the plane normal to the 180° rotation axis. It follows that a twin orientation always implies the existence of a CSL, but the reverse is not always true[59,5]; for example, in the case of $\Sigma = 39$, there is no axis-angle pair representation with an angle of rotation of 180° , so that it is not possible to find a coherent interface between crystals related by a Σ = 39 orientation relation.

(ii) Boundaries containing a high absolute density of coincidence sites (i.e., a large value of number of coincidence sites per unit boundary area) can in general be expected to have the lowest energy.

(iii) Calculations of atomic positions [60-62] in the boundary region, using interatomic force laws, suggest that in materials where the atoms are hard (strong repulsive interaction at short range), coincidence site lattices <u>may</u> not exist. For example, in the case of a $\Sigma 3$ twin in a "hard" BCC material, with a {1 1 2} coherent twin plane, it is found that a small rigid translation (by a vector a/12<$\bar{1}$ $\bar{1}$ 1>) of the twin lattice lowers the energy of the interface (Fig. 24)[27,5]. Because of this relaxation, the lattices no longer have a common origin and so the coincidences vanish. Nevertheless, boundaries which contain high densities of coincidence sites before relaxation may be expected to represent better fit between the lattices, and thus have low energies relative to other boundaries. This is because the periodic nature and the actual repeat period of the structure of the interface implied by the CSL concept is not destroyed by the small translation. The rigid body translations mentioned above have been experimentally established in the case of Aluminium; although not conclusively established, the experiments suggest that the translation may have a component outside the interface plane, but the atomistic calculations cannot predict this since they always seem to be carried out at constant volume [27].

(iv) The physical significance of CSL's must diminish as the Σ value increases, because only a very small fraction of atoms in an interface can then be common to both the adjacent crystals.

(v) The Σ value is independent [5] of the transformation matrix (eg. (A J A) in example 25) and hence is independent of symmetry considerations. This is convenient since such matrices do not uniquely relate the grains; it is usually necessary to impose criteria to allow physically reasonable choices of (A J A) to be made.

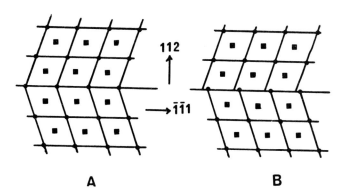

112

$\rightarrow \bar{1}\bar{1}1$

A   **B**

<u>Fig. 24</u>:   {1 1 2} coherent twin boundary in a BCC material, initially with an exact $\Sigma 3$ CSL. The figure on the right illustrates the structure after a rigid body translation is included [27,5].

81

The O-lattice

We have seen that coincidence site lattices can be generated by identifying lattice points which are common to both of the adjacent crystals, when the lattices of these crystals are notionally allowed to interpenetrate and fill all space. The totality of the common lattice points then forms the coincidence site lattice. This procedure is of course only a formal way of identifying coincidence points, and for a real interface connecting the two crystals, only atoms from crystal 1 will exist on one side of the interface and those from crystal 2 on the other. There will, however, be good fit regions in the interface, corresponding to the lattice points that are common to both crystals; the pattern formed by these points consists of a two dimensional section of the CSL. The coincidence site lattice thus concentrates on just lattice points, and this is not the most general case to consider. When the transformation carrying one grain into the other is a rigid body rotation, all points along the rotation axis will represent points of perfect fit between the two crystals, not just the lattice points which may lie on the rotation axis. When the transformation relating two grains is an invariant-plane strain, all points in the invariant-plane represent points of perfect fit between the two lattices and not just the lattice points which lie in the invariant-plane.

Any point $\underline{x}$ within a crystal may be represented as the sum of a lattice vector $\underline{u}$ (which has integral components) and a small vector β whose components are fractional and less than unity. The internal co-ordinates of the point $\underline{x}$ are then defined to be the components of the vector β. The O-lattice method takes account of all coincidences, between non-lattice sites of identical internal co-ordinates as well as the coincidence lattice sites.

All lattice points in a crystal are crystallographically equivalent and any lattice point may be used as an origin to generate the three dimensional crystal lattice. To identify points of the CSL we specify that any lattice vector $\underline{u}$ of crystal A must, as a result of the transformation to crystal B, become another lattice vector $\underline{w}$ of A - i.e., $\underline{w} = \underline{u}+\underline{v}$ where $\underline{v}$ is a lattice vector of A. The CSL point $\underline{w}$ can then be considered to be a perfect fit point in an interface between A and B, because it corresponds to a lattice point in both crystals.

Non-lattice points in a crystal are crystallographically equivalent when they have the same internal co-ordinates. To identify O-points [64,65], we specify that any non-lattice vector $\underline{x}$ of crystal A must, as a result of transformation to crystal B, become another non-lattice vector $\underline{y}$ of A such that $\underline{y} = \underline{x}+\underline{v}$ where $\underline{v}$ is a lattice vector of A; the points $\underline{x}$ and $\underline{y}$ thus have the same internal co-ordinates in A. The O-point $\underline{y}$ can then be considered to be a perfect fit point in the interface between A and B. Note that when $\underline{x}$ becomes a lattice vector, $\underline{y}$ becomes a CSL point. The totality of O-points obtained by allowing crystals A and B to notionally interpenetrate forms the O-lattice [64,65], which may contain the CSL as a sub-lattice if A and B are suitably orientated at an exact CSL orientation. Any boundary between A and B cuts the O-lattice and will contain regions of good fit which have the periodicity corresponding to the periodicity of a planar net of the O-lattice along which the intersection occurs. Boundaries parallel to low-index planes of the O-lattice are in general two dimensionally periodic with relatively small repeat cells, and those with a high planar O-point density should have a relatively low energy.

Consider two crystals A and B which are related by the deformation (A S A) which converts the reference lattice A to that of B; an arbitrary non-lattice point $\underline{x}$ in crystal A thus becomes a point $\underline{y}$ in crystal B, where

$$[A;\underline{y}] = (A\ S\ A)[A;\underline{x}]$$

If the point $\underline{y}$ is crystallographically equivalent to the point $\underline{x}$, in the sense that it has the same

82

internal co-ordinates as $\underline{x}$, then $\underline{y}$ is also a point of the O-lattice, designated $\underline{O}$. This means that $\underline{y} = \underline{x}+\underline{u}$, where $\underline{u}$ is a lattice vector of A. Since $\underline{y}$ is only an O-point when $\underline{y} = \underline{x}+\underline{u}$, we may write that $\underline{y}$ is an O-lattice vector $\underline{O}$ if [(64,65)]

$$[A;\underline{y}] = [A;\underline{O}] = [A;\underline{x}] + [A;\underline{u}] = (A\ S\ A)[A;\underline{x}]$$

or in other words,

$$[A;\underline{u}] = (A\ T\ A)[A;\underline{O}] \qquad \qquad .....(30a)$$

where $(A\ T\ A) = I - (A\ S\ A)^{-1}$.

It follows that

$$[A;\underline{O}] = (A\ T\ A)^{-1}[A;\underline{u}] \qquad \qquad .....(30b)$$

By substituting for $\underline{u}$ the three basis vectors of A in turn [(5)], we see that the columns of $(A\ T\ A)^{-1}$ define the corresponding base vectors of the O-lattice.

Since O-points are points of perfect fit, mismatch must be at a maximum in between neighbouring O-points. When $\underline{O}$ is a primitive O-lattice vector, eq.30b states that the amount of misfit in between the two O-points connected by $\underline{O}$ is given by $\underline{u}$. A dislocation with Burgers vector $\underline{u}$ would thus accommodate this misfit and localise it at a position between the O-points, and these ideas allow us to consider a dislocation model of the interface in terms of the O-lattice theory.

Three sets of dislocations with Burgers vectors $\underline{b}_i$ (which form a non-coplanar set) are in general required to accommodate the misfit in any interface. If the Burgers vectors $\underline{b}_1$, $\underline{b}_2$ and $\underline{b}_3$ are chosen to serve this purpose, and each in turn substituted into eq.30b, then the corresponding O-lattice vectors $\underline{O}_1$, $\underline{O}_2$ and $\underline{O}_3$ are obtained. These vectors $\underline{O}_i$ thus define the basis vectors of an O-lattice unit cell appropriate to the choice of $\underline{b}_i$. If the O-points of this O-lattice are separated by "cell walls" which bisect the lines connecting neighbouring O-points, then the accumulating misfit in any direction can be considered to be concentrated at these cell walls [(64,65)]. When a real interface (unit normal $\underline{n}$) is introduced into the O-lattice, its line intersections with the cell walls become the interface dislocations with Burgers vectors $\underline{b}_i$ and unit line vectors $\underline{l}_i$ parallel to the line of intersection of the interface with the cell walls.

The three O-lattice cell walls (with normals $\underline{O}_i^*$) have normals parallel to $\underline{O}_1^* = \underline{O}_2 \wedge \underline{O}_3$, $\underline{O}_2^* = \underline{O}_3 \wedge \underline{O}_1$ and $\underline{O}_3^* = \underline{O}_1 \wedge \underline{O}_2$, so that the line vectors of the dislocations are given by $\underline{l}_1 \| \underline{O}_1^* \wedge \underline{n}$, $\underline{l}_2 \| \underline{O}_2^* \wedge \underline{n}$ and $\underline{l}_3 \| \underline{O}_3^* \wedge \underline{n}$. Similarly, $1/d_i = |\underline{O}_i^* \wedge \underline{n}|$. These results are exactly equivalent to the theory developed at the beginning of this chapter (eq.25-27) but the O-lattice theory perhaps gives a better physical picture of the interface, and follows naturally from the CSL approach [(5)]. The equivalence of the two approaches arises because eq.30a is identical to eq.25 since $\underline{u}$ and $\underline{O}$ are equivalent to $\underline{b}_t$ and $\underline{p}$ respectively.

Example 26: The alpha/beta brass interface using O-lattice theory

Derive the structure of the alpha/beta brass interface of example 23 using O-lattice theory.
From example 23, the matrix $(A\ T\ A)$ is given by:

$$(A\ T\ A) = \begin{pmatrix} 0.119500 & 0.006386 & 0.016048 \\ 0.006386 & 0.119500 & 0.016048 \\ -0.149974 & -0.149974 & -0.236183 \end{pmatrix}$$

so that

$$(A \ T \ A)^{-1} = \begin{pmatrix} -52.448 & -61.289 & -51.069 \\ -61.289 & -52.448 & -51.064 \\ 72.222 & 72.222 & 60.622 \end{pmatrix}$$

If the vectors $\underline{b}_1$, $\underline{b}_2$ and $\underline{b}_3$ are each in turn substituted for $\underline{u}$ in eq.30a, then the corresponding $\underline{O}$ lattice vectors are found to be:

$$[A;\underline{O}_1] = [-0.689770 \quad -5.110092 \quad 5.799861]$$
$$[A;\underline{O}_2] = [-5.110092 \quad -0.689770 \quad 5.799861]$$
$$[A;\underline{O}_3] = [-56.86861 \quad -56.86861 \quad 72.22212]$$

$$[A^*;\underline{O}_1^*] = [\ 0.263088 \quad 0.036860 \quad 0.236183]$$
$$[A^*;\underline{O}_2^*] = [\ 0.036860 \quad 0.263088 \quad 0.236183]$$
$$[A^*;\underline{O}_3^*] = [-0.024088 \quad -0.024088 \quad -0.024088]$$

$$\underline{O}_1^* \wedge \underline{n}\text{III}_1 \ = [-0.012249 \quad -0.012249 \quad 0.015556]_A$$
$$\underline{O}_2^* \wedge \underline{n}\text{III}_2 \ = [-0.012249 \quad -0.012249 \quad 0.015556]_A$$
$$\underline{O}_3^* \wedge \underline{n}\text{III}_3 \ = [\ 0.001249 \quad 0.001249 \quad -0.002498]_A$$

$$d_1 \ = d_2 \ = 11.63A^O \quad \text{and} \ d_3 \ = 88.52A^O$$

The results obtained are therefore identical to those of example 23.

Secondary Dislocations

Low-angle boundaries contain interface dislocations whose role is to localise and accommodate the misfit between adjacent grains, such that large areas of the boundary consist of low-energy coherent patches without mismatch. These intrinsic interface dislocations are called *primary* dislocations because they accommodate the misfit relative to an ideal single crystal as the reference lattice.

Boundaries between grains which are at an exact CSL orientation have relatively low-energy. It then seems reasonable to assume that any small deviation from the CSL orientation should be accommodated by a set of interface dislocations which localise the misfit due to this deviation, and hence allow the perfect CSL to exist over most of the boundary area. These intrinsic interface dislocations [66-69] are called *secondary* dislocations because they accommodate the misfit relative to a CSL as the reference lattice. High-angle boundaries between crystals which are not at an exact CSL orientation may therefore consist of dense arrays of primary dislocations and also relatively widely spaced arrays of secondary dislocations. The primary dislocations may be so closely spaced that their strain fields virtually cancel each other and in these circumstances only secondary dislocations would be visible using conventional transmission electron microscopy.

Example 27: Intrinsic secondary dislocations

The axis-angle pair describing the orientation relationship between the two grains (A and B) of austenite is given by:

axis of rotation parallel to $[1\ 1\ 2]_A$

right-handed angle of rotation 175°

Calculate the secondary dislocation structure of an interface lying normal to the axis of rotation, given that the Burgers vectors of the interface dislocations are $[A;\underline{b}_1] = 0.5[1\ 0\ \bar{1}]$, $[A;\underline{b}_2] = 0.5[0\ 1\ \bar{1}]$ and $[A;\underline{b}_3] = 0.5[1\ 1\ 0]$, where the basis A corresponds to the conventional FCC unit cell of austenite, with a lattice parameter a = 3.56 angstroms.

The secondary dislocation structure can be calculated with respect to the nearest CSL, which is a $\Sigma 3$ CSL obtained by a 180° rotation about $[1\ 1\ 2]_A$. The rigid body rotation matrix corresponding to this exact CSL orientation is thus:

$$(A\ J\ A) = (1/3) \begin{pmatrix} \bar{2} & 1 & 2 \\ 1 & \bar{2} & 2 \\ 2 & 2 & 1 \end{pmatrix}$$

The rotation matrix (A J2 A) describing the actual transformation of A to B, corresponding to a rotation of 175° about $[1\ 1\ 2]_A$ is given by:

$$(A\ J2\ A) = \begin{pmatrix} -0.663496 & 0.403861 & 0.629817 \\ 0.261537 & -0.663496 & 0.700979 \\ 0.700979 & 0.629817 & 0.334602 \end{pmatrix}$$

The matrix (A J3 A) describing the deviation from the exact CSL is given by (65)
(A J3 A) = (A J2 A)$^{-1}$(A J A), so that

$$(A\ J3\ A) = \begin{pmatrix} 0.996829 & 0.071797 & -0.034313 \\ -0.070528 & 0.996829 & 0.036850 \\ 0.036850 & -0.034313 & 0.998732 \end{pmatrix}$$

If (A T A) = I - (A J3 A)$^{-1}$, then (A T' A) is given by:

$$(A\ T'\ A) = \begin{pmatrix} 0.003171 & -0.071797 & 0.034313 \\ 0.070528 & 0.003171 & -0.036850 \\ -0.036850 & 0.034313 & 0.001268 \end{pmatrix}$$

The secondary dislocation structure can now be calculated using eq.26f,27c,27d:

$$(\underline{n};A^*) = a(0.408248\ 0.408248\ 0.816497)$$
$$[A^*;\underline{b}_1^*] = [1\ \bar{1}\ 1]$$
$$[A^*;\underline{b}_2^*] = [\bar{1}\ 1\ \bar{1}]$$
$$[A^*;\underline{b}_3^*] = [1\ 1\ 1]$$

From eq.26f,

$$[A^*;\underline{c}_1] = [0.040655\ 0.104207\ -0.072341]$$
$$[A^*;\underline{c}_2] = [-0.0109280\ -0.030507\ 0.069894]$$

$$[A^*;\underline{c}_3] = [-0.034313 \quad 0.036850 \quad -0.001268]$$

so that

$$d_1 = 26.71 \; A^o, \; d_2 = 26.71 \; A^o \quad \text{and} \; d_3 = 70.68 \; A^o$$
$$\underline{l}_1 \; \|[\; 0.860386 \quad -0.470989 \quad -0.194695]_A$$
$$\underline{l}_2 \; \|[-0.401046 \quad 0.883699 \quad -0.241327]_A$$
$$\underline{l}_3 \; \|[\; 0.607647 \quad 0.545961 \quad -0.576794]_A$$

The DSC lattice

In a boundary between two crystals, the Burgers vector $\underline{b}$ of an interface dislocation must be such that the displacement of one of the crystals through $\underline{b}$ relative to the other does not change the structure of the interface. Lattice translation vectors of the reference lattice always satisfy this condition, so that $\underline{b}$ can always equal a lattice translation vector. However, additional possibilities arise in the case of secondary dislocations, whose Burgers vectors are generally DSC lattice translation vectors[70,65].

Secondary dislocations represent the deviation from a particular coincidence site lattice, and a corresponding DSC lattice may be generated such that the translation vectors of the DSC lattice are possible Burgers vectors of secondary interface dislocations. The interesting point about the DSC lattice is that its lattice vectors need not be crystal lattice translation vectors. Fig. 25 illustrates a $\Sigma 5$ CSL between two FCC grains (A & B), related by a rotation of 36.87^o about $[1 \; 0 \; 0]_A$.

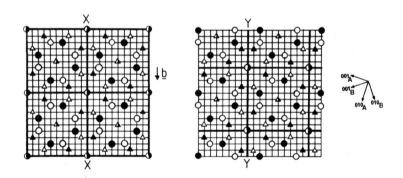

Fig. 25: $\Sigma 5$ coincidence system for FCC crystals [71]. Filled symbols are lattice A, unfilled ones lattice B and coincidence sites are a mixture of the two; lattice sites in the plane of the diagram are represented as circles whereas those displaced by $(1/2)[1 \; 0 \; 0]$ are represented as triangles. The $[1 \; 0 \; 0]$ axis is normal to the plane of the diagram.

Fig. 25b is obtained by displacing lattice B by $[A;\underline{b}] = (1/10)[0 \; 3 \; 1]$ relative to lattice A and it is obvious that the basic pattern of lattice sites and CSL sites remains unaffected by this translation, despite the fact that $\underline{b}$ is not a lattice vector of A or B. It is thus possible for secondary dislocations to have Burgers vectors which are not lattice translation vectors, but are vectors of the

DSC lattice. The DSC lattice, or the Displacement Shift Complete lattice, is the coarsest lattice which contains the lattice points of both A and B, and any DSC lattice vector is a possible Burgers vector for a perfect secondary dislocation. We note that the displacement $\underline{b}$ causes the original coincidences (Fig. 25a) to disappear and be replaced by an equivalent set of new coincidences (Fig. 25b), and this always happens when $\underline{b}$ is not a lattice translation vector. This shift of the origin of the CSL has an important consequence on the topography [71] of any boundary containing secondary dislocations with non-lattice Burgers vectors.

Considering again Fig. 25, suppose that we introduce a boundary into the CSL, with unit normal [A;$\underline{n}$]‖[0 1 $\overline{2}$], so that its trace is given by XX on Fig. 25a. The effect of the displacement $\underline{b}$ of crystal B relative to A, due to the presence of a secondary dislocation, is to shift the origin of the CSL; if the boundary originally at XX is to have the same structure after the displacement then it has to shift to the position YY in Fig. 25b. Because a dislocation separates slipped from unslipped regions, the shift of the boundary occurs at the position of the secondary dislocation so that the boundary is stepped at the core of this dislocation. One such step is illustrated in Fig. 26.

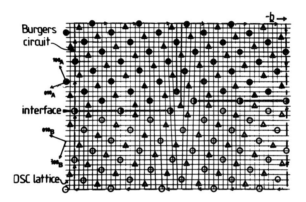

Fig. 26: The presence of a step [72] in a Σ5, (3 $\overline{1}$ 0)$_A$ boundary of an FCC crystal containing a secondary interface dislocation with [A;$\underline{b}$] = (a/10)[1 3 0]. The symbolism is identical to that of Fig. 25.

The following further points about the DSC lattice and its consequences should be noted:

(i) The DSC lattice can be constructed graphically simply by inspection, bearing in mind that it is the coarsest lattice containing lattice sites from both the crystals orientated at an exact CSL orientation. Rather detailed analytical methods for computing the basis vectors of the DSC lattice have been presented elsewhere [73], and tabulations of these DSC lattice calculations as a function of Σ can also be found for cubic systems [74].

(ii) For the primitive cubic system, the components of the three basis vectors of the DSC lattice are the columns of a 3x3 matrix (DSC) obtained by taking the transpose of the inverse of another 3x3 matrix (CSL). The columns of (CSL) represent the components of the basis vectors of the coincidence site lattice concerned.

(iii) Crystal lattice vectors (of both the adjacent crystals) form a sub-set of DSC lattice vectors.

(iv) The volume of a CSL unit cell is Σ times larger than that of the crystal lattice unit cell, whereas the DSC lattice unit cell has a volume $1/\Sigma$ times that of the crystal lattice unit cell [73].

(v) The homogeneous deformation (A S A) of eq.25, describing the transformation from the reference lattice to another crystal is not unique [5], but the CSL and DSC concepts are independent of the choice of (A S A). The O-lattice on the other hand, depends critically on the form of (A S A).

(vi) Primitive DSC lattice vectors can be much smaller than primitive crystal lattice vectors [75]. On the basis of elastic strain energy arguments, smaller interface dislocation Burgers vectors should be favoured. This is not confirmed experimentally since secondary dislocations often have Burgers vectors which are crystal lattice vectors.

(vii) Although DSC lattices are defined relative to the unrelaxed CSL, small rigid body translations which destroy exact coincidence (but preserve the CSL periodicity) do not affect the essentials of the DSC concept [27].

Some difficulties associated with interface theory

Consider two crystals A and B, both of cubic structure, related by a rigid body rotation. Any boundary containing the axis of rotation is a tilt boundary; for the special case of the symmetrical tilt boundary, lattice B can be generated from A by reflection across the boundary plane. By substituting the rigid body rotation for (A S A) in eq.25, the dislocation structure of the symmetrical tilt boundary may be deduced (example 22) to consist of a single array of dislocations with line vectors parallel to the tilt axis.

Symmetry considerations imply that the rigid body rotation has up to 23 further axis-angle representations. If we impose the condition that the physically most significant representation is that which minimizes the Burgers vector content of the interface, then the choice reduces to the axis-angle pair involving the smallest angle of rotation.

On the other hand, (A S A) can also be a lattice-invariant twinning shear on the symmetrical tilt boundary plane [5]; crystal B would then be related to A by reflection across the twin plane so that the resulting bicrystal would be equivalent to the case considered above. The dislocation content of the interface then reduces to zero since the invariant-plane of the twinning shear is fully coherent.

This ambiguity in the choice of (A S A) is a major difficulty in interface theory [5]. The problem is compounded by the fact that interface theory is phenomenological - i.e., the transformation strain (A S A) may be real or notional as far as interface theory is concerned. If it is real then we expect to observe a change in the shape of the transformed region, and this may help in choosing the most reasonable deformation (A S A). For example, in the case of mechanical twinning in FCC crystals, the surface relief observed can be used to deduce that (A S A) is a twinning shear rather than a rigid body rotation. In the case of an FCC annealing twin, which grows from the matrix by a diffusional mechanism (during grain boundary migration), the same twinning shear (A S A) may be used to deduce the interface structure, but the deformation is now notional, since the formation of annealing twins is not accompanied by any surface relief effects. In these circumstances, we cannot be certain that the deduced interface structure for the annealing twin is correct.

The second major problem follows from the fact that the mathematical Burgers vector content $\underline{b}_t$ given by eq.25 has to be factorised into arrays of physically realistic dislocations with Burgers vectors which are vectors of the DSC lattice. There is an infinite number of ways in which this can be done, particularly since the interface dislocations do not necessarily have Burgers vectors which minimise their elastic strain energy.

Secondary dislocations are referred to an exact CSL as the reference lattice. Atomistic calculations [76] suggest that boundaries in crystals orientated at exact coincidence contain arrays of primary dislocations (whose cores are called "structural units"). The nature of the structural units varies with the Σ value, but some "favoured" CSL's have boundaries with just one type of structural unit, so that the stress field of the boundary is very uniform. Other CSL's have boundaries consisting of a mixture of structural units from various favoured CSL's. It has been suggested that it is the favoured CSL's which should be used as the reference lattices in the calculation of secondary dislocation structure, but the situation is unsatisfactory because the same calculations suggest that the favoured/unfavoured status of a boundary also depends of the boundary orientation itself. (We note that the term "favoured" does not imply a low interface energy).

Referring now to the rigid body translations which exist in materials with "hard" atoms, it is not clear whether the calculated translations might be different if the relaxation were to be carried out for a three dimensionally enclosed particle.

APPENDIX 1: VECTORS AND MATRICES

Vectors

Quantities (such as force, displacement) which are characterised by both magnitude and direction are called vectors; scalar quantities (such as time) only have magnitude. A vector is represented by an arrow pointing in a particular direction, and can be identified by underlining the lower-case vector symbol (e.g., $\underline{u}$). The magnitude of $\underline{u}$ (or $|\underline{u}|$) is given by its length, a scalar quantity. Vectors $\underline{u}$ and $\underline{v}$ are only equal if they both point in the same direction, and if $|\underline{u}| = |\underline{v}|$. The parallelism of $\underline{u}$ and $\underline{v}$ is indicated by writing $\underline{u}||\underline{v}$. If $\underline{w} = -\underline{u}$, then $\underline{w}$ points in the opposite direction to $\underline{u}$, although $|\underline{w}| = |\underline{u}|$.

Vectors can be added or removed to give new vectors, and the order in which these operations are carried out is not important. Vectors $\underline{u}$ and $\underline{x}$ can be added by placing the initial point of $\underline{x}$ in contact with the final point of $\underline{u}$; the initial point of the resultant vector $\underline{u}+\underline{x}$ is then the initial point of $\underline{u}$ and its final point corresponds to the final point of $\underline{x}$. The vector $m\underline{u}$ points in the direction of $\underline{u}$, but $|m\underline{u}|/|\underline{u}| = m$, m being a scalar quantity. A unit vector has a magnitude of unity; dividing a vector $\underline{u}$ by its own magnitude u gives a unit vector parallel to $\underline{u}$.

It is useful to refer vectors to a fixed frame of reference; an arbitrary reference frame would consist of three non-coplanar basis vectors $\underline{a}_1$, $\underline{a}_2$ and $\underline{a}_3$. The vector $\underline{u}$ could then be described by means of its components u_1, u_2 and u_3 along these basis vectors, respectively, such that

$$u_i = |\underline{u}|\cos(\theta_i)/|\underline{a}_i| \quad \text{where } i = 1,2,3 \text{ and } \theta_i = \text{angle between } \underline{u} \text{ and } \underline{a}_i.$$

If the basis vectors $\underline{a}_i$ form an orthonormal set (i.e., they are mutually perpendicular and each of unit magnitude), then the magnitude of $\underline{u}$ is:

$$|\underline{u}|^2 = u_1^2 + u_2^2 + u_3^2$$

If the basis vectors $\underline{a}_i$ form an orthogonal set (i.e., they are mutually perpendicular) then the magnitude of $\underline{u}$ is:

$$|\underline{u}|^2 = (u_1|\underline{a}_1|)^2 + (u_2|\underline{a}_2|)^2 + (u_3|\underline{a}_3|)^2$$

A 'dot' or 'scalar' product between two vectors $\underline{u}$ and $\underline{x}$ (order of multiplication not important) is given by $\underline{u}.\underline{x} = |\underline{u}|\,|\underline{x}|\cos\theta$, θ being the angle between $\underline{u}$ and $\underline{x}$. If $\underline{x}$ is a unit vector then $\underline{u}.\underline{x}$ gives the projection of $\underline{u}$ in the direction $\underline{x}$.

The 'cross' or 'vector' product is written $\underline{u} \wedge \underline{x} = |\underline{u}|\,|\underline{x}|\sin\theta\;\underline{y}$, where $\underline{y}$ is a unit vector perpendicular to both $\underline{u}$ and $\underline{x}$, with $\underline{u},\underline{x}$ and $\underline{y}$ forming a right-handed set. A right-handed set $\underline{u}$, $\underline{x}$, $\underline{y}$ implies that a right-handed screw rotated through an angle less than 180° from $\underline{u}$ to $\underline{x}$ advances in the direction $\underline{y}$. The magnitude of $\underline{u} \wedge \underline{x}$ gives the area enclosed by a parallelogram whose sides are the vectors $\underline{u}$ and $\underline{x}$; the vector $\underline{y}$ is normal to this parallelogram. Clearly, $\underline{u} \wedge \underline{x} \neq \underline{x} \wedge \underline{u}$.

If $\underline{u},\underline{x}$ and $\underline{z}$ form a right-handed set of three non-coplanar vectors then $\underline{u} \wedge \underline{x}.\underline{z}$ gives the volume of the parallelopiped formed by $\underline{u}$, $\underline{x}$ and $\underline{z}$. It follows that $\underline{u} \wedge \underline{x}.\underline{z} = \underline{u}.\underline{x} \wedge \underline{z} = \underline{z} \wedge \underline{u}.\underline{x}$.

The following relations should be noted:

$$\underline{u} \wedge \underline{x} = -\underline{x} \wedge \underline{u}$$
$$\underline{u}.(\underline{x} \wedge \underline{y}) = \underline{x}.(\underline{y} \wedge \underline{u}) = \underline{y}.(\underline{u} \wedge \underline{x})$$
$$\underline{u} \wedge (\underline{x} \wedge \underline{y}) \neq (\underline{u} \wedge \underline{x}) \wedge \underline{y}$$
$$\underline{u} \wedge (\underline{x} \wedge \underline{y}) = (\underline{u}.\underline{y})\underline{x} - (\underline{u}.\underline{x})\underline{y}$$

Matrices

Definition, addition, scalar multiplication

A matrix is a rectangular array of numbers, having m rows and n columns, and is said to have an order "m by n". A square matrix $\underline{J}$ of order 3 by 3 may be written as

$$\underline{J} = \begin{pmatrix} J_{11} & J_{12} & J_{13} \\ J_{21} & J_{22} & J_{23} \\ J_{31} & J_{32} & J_{33} \end{pmatrix} \quad \text{and its transpose} \quad \underline{J}' = \begin{pmatrix} J_{11} & J_{21} & J_{31} \\ J_{12} & J_{22} & J_{32} \\ J_{13} & J_{23} & J_{33} \end{pmatrix}$$

where each number J_{ij} (i = 1,2,3 and j = 1,2,3) is an element of $\underline{J}$. $\underline{J}'$ is called the transpose of the matrix $\underline{J}$. An identity matrix $(\underline{I})$ has the diagonal elements J_{11}, J_{22} & J_{33} equal to unity, all the other elements being zero. The trace of a matrix is the sum of all its diagonal elements $(J_{11}+J_{22}+J_{33})$. If matrices $\underline{J}$ and $\underline{K}$ are of the same order, they are said to be equal when $J_{ij} = K_{ij}$ for all i,j. Multiplying a matrix by a constant involves the multiplication of every element of that matrix by that constant. Matrices of the same order may be added or subtracted, so that if $\underline{L} = \underline{J} + \underline{K}$, it follows that $L_{ij} = J_{ij} + K_{ij}$.

The Einstein Summation Convention

In order to simplify more complex matrix operations we now introduce the summation convention. The expression

$$u_1 a_1 + u_2 a_2 + ... u_n a_n$$

can be shortened by writing

$$\sum_{i=1}^{n} u_i a_i$$

A further economy in writing is achieved by adopting the convention that the repetition of a subscript or superscript index in a given term implies summation over that index from 1 to n. Using this summation convention, the above sum can be written $u_i a_i$. Similarly,

$$x_i = y_j z_{ij} \quad \text{for i = 1,2 and j = 1,2 implies that}$$
$$x_1 = y_1 z_{11} + y_2 z_{12}$$
$$x_2 = y_1 z_{21} + y_2 z_{22}$$

Multiplication and Inversion

The matrices $\underline{J}$ and $\underline{K}$ can be multiplied in that order to give a third matrix $\underline{L}$ if the number of columns (m) of $\underline{J}$ equals the number of rows of $\underline{K}$ ($\underline{J}$ is said to be conformable to $\underline{K}$). $\underline{L}$ is given by

$$L_{st} = J_{sr}K_{rt}$$

where s ranges from 1 to the total number of rows in $\underline{J}$ and t ranges from 1 to the total number of columns in $\underline{K}$. If $\underline{J}$ and $\underline{K}$ are both of order 3x3 then

for example, $$L_{11} = J_{11}K_{11}+J_{12}K_{21}+J_{13}+K_{31}$$

Note that the product $\underline{JK}$ does not in general equal $\underline{KJ}$.

Considering a nxn square matrix $\underline{J}$, it is possible to define a number Δ which is the determinant (of order n) of $\underline{J}$. A *minor* of any element J_{ij} is obtained by forming a new determinant of order (n-1), of the matrix obtained by removing all the elements in the ith row and the jth column of $\underline{J}$. For example, if $\underline{J}$ is a 2x2 matrix, the minor of J_{11} is simply J_{22}. If $\underline{J}$ is a 3x3 matrix, the minor of J_{11} is:

$$\begin{vmatrix} J_{22} & J_{23} \\ J_{32} & J_{33} \end{vmatrix} = J_{22}J_{33} - J_{23}J_{32}$$

where the vertical lines imply a determinant. The cofactor j_{ij} of the element J_{ij} is then given by multiplying the minor of J_{ij} by $(-1)^{i+j}$. The determinant (Δ) of $\underline{J}$ is thus

$$\det\underline{J} = \sum_{j=1}^{n} J_{1j}j_{1j} \qquad \text{with } j = 1,2,3$$

Hence, when $\underline{J}$ is a 3x3 matrix, its determinant Δ is given by:

$$\Delta = J_{11}j_{11} + J_{12}j_{12}+J_{13}j_{13}$$

$$= J_{11}(J_{22}J_{33}-J_{23}J_{32})+J_{12}(J_{23}J_{31}-J_{21}J_{33})+J_{13}(J_{21}J_{32}-J_{22}J_{31})$$

The inverse of $\underline{J}$ is written $\underline{J}^{-1}$ and is defined such that the $\underline{J}.\underline{J}^{-1} = \underline{I}$. The elements of $\underline{J}^{-1}$ are J^{-1}_{ij} such that:

$$J^{-1}_{ij} = j_{ji}/\det\underline{J}$$

Hence, if $\underline{L}$ is the inverse of $\underline{J}$, and if $\det\underline{J} = \Delta$, then:

$$L_{11} = (J_{22}J_{33}-J_{23}J_{32})/\Delta$$
$$L_{12} = (J_{32}J_{13}-J_{33}J_{12})/\Delta$$
$$L_{13} = (J_{12}J_{23}-J_{13}J_{22})/\Delta$$

$$L_{21} = (J_{23}J_{31}-J_{21}J_{33})/\Delta$$
$$L_{22} = (J_{33}J_{11}-J_{31}J_{13})/\Delta$$
$$L_{23} = (J_{13}J_{21}-J_{11}J_{23})/\Delta$$

$$L_{31} = (J_{21}J_{32}-J_{22}J_{31})/\Delta$$
$$L_{32} = (J_{31}J_{12}-J_{32}J_{11})/\Delta$$
$$L_{33} = (J_{11}J_{22}-J_{12}J_{21})/\Delta$$

If the determinant of a matrix is zero, the matrix is singular and does not have an inverse. The following matrix equations are noteworthy:

$$(\underline{J}\ \underline{K})\ \underline{L} = \underline{J}\ (\underline{K}\ \underline{L})$$
$$\underline{J}\ (\underline{K} + \underline{L}) = \underline{J}\ \underline{K} + \underline{J}\ \underline{L}$$
$$(\underline{J} + \underline{K})' = \underline{J}' + \underline{K}'$$
$$(\underline{J}\ \underline{K})' = \underline{K}'\ \underline{J}'$$
$$(\underline{J}\underline{K})^{-1} = \underline{K}^{-1}\underline{J}^{-1}$$
$$(\underline{J}^{-1})' = (\underline{J}')^{-1}$$

Orthogonal Matrices

A square matrix $\underline{J}$ is said to be orthogonal if $\underline{J}^{-1} = \underline{J}'$. It the columns or rows of a real orthogonal matrix are taken to be components of column or row vectors respectively, then these vectors are all unit vectors. The set of column vectors (or row vectors) form an orthonormal basis; if this basis is right-handed, the determinant of the matrix is unity, but if it is left-handed then $\Delta = -1$.

Orthogonal matrices arise in co-ordinate transformations between orthonormal bases and where rigid body rotations are represented in a single orthonormal basis. References 77, 78 should be consulted for further information on matrices and vectors.

REFERENCES

1) M.S. Wechsler, D.S. Lieberman and T.A. Read, Trans. Amer. Inst. Min. Metall. Engrs., 1953, 197, 1503.
2) J.S. Bowles and J.K. MacKenzie, Acta Metall., 1954, 2, 129.
3) E.C. Bain, Trans. Amer. Inst. Min. Metall. Engrs., 1924, 70, 25.
4) J.W. Christian, "The Theory of Transformations in Metals and Alloys", 1965, Oxford, Pergamon Press.
5) J.W. Christian, "The Theory of Transformations in Metals and Alloys", Part 1, 2nd Edition, 1975, Oxford, Pergamon Press.
6) P.L. Ryder and W. Pitsch, Acta Metall., 1966, 14, 1437.
7) P.L. Ryder, W. Pitsch and R.F. Mehl, Acta Metall., 1967, 15, 1431.
8) J.W. Christian, "The Mechanism of Phase Transformations in Crystalline Solids", Inst. of Metals Monograph 33, 1969, p.129, London.
9) Yu. A. Bagaryatski, Dokl. Akad. Nauk, SSSR, 1950, 73, 1161.
10) K.W. Andrews, Acta Metall., 1963, 11, 939.
11) W. Hume-Rothery, G.V. Raynor and A.T. Little, Arch. Eisenhuttenw., 1942, 145, 143.
12) D.N. Shackleton and P.M. Kelly, Acta Metall., 1967, 15, 979.
13) H.K.D.H. Bhadeshia, Acta Metall., 1980, 28, 1103.
14) W. Pitsch, Acta Metall., 1962, 10, 897.
15) G.V. Kurdjumov and G. Sachs, Z. Phys., 1930, 64, 325.
16) Z. Nishiyama, Sci. Rept. Tohoku Univ., 1934, 23, 325.
17) A. Crosky, P.G. McDougall and J.S. Bowles, Acta Metall., 1980, 28, 1495.
18) F.F. Lange, Acta Metall., 1967, 15, 311; C. Goux, Cand. Metall. Quarterly, 1974, 13 9.
19) P.C. Rowlands, E.O. Fearon and M. Bevis, Trans. AIME, 1968, 242, 1559.
20) Y.C. Liu, Trans. AIME, 1963, 227, 775.
21) J.W. Brooks, M.H. Loretto and R.E. Smallman, Acta Metall., 1979, 27, 1839.
22) C.M. Wayman, "Introduction to the Crystallography of Martensitic Transformations", Macmillan, New York, 1964.
23) E. Schmid and G. Boas, "Kristallplastizitat", Springer, Berlin, 1936.
24) J.W. Christian, Metall. Trans. A, 1982, 13A, 509.
25) A. Kelly and G.W. Groves, "Crystallography and Crystal Defects", Longmans, London, 1970.
26) D.K. Bowen and J.W. Christian, Phil. Mag., 1965, 12, 369.
27) J.W. Christian and A.G. Crocker, "Dislocations in Solids", North Holland, Amsterdam, 1980, 3, 165 (Ed. F.R.N. Nabarro).
28) J.D. Eshelby, Proc. Roy. Soc., 1957, A241, 376.
29) J.W. Christian, Acta Metall., 1958, 6, 377.
30) J.W. Christian, Proc. Int. Conf. on Martensitic Transformations., ICOMAT 1979, Massachusetts, p.220.
31) J.W. Christian, 'Strengthening methods in crystals", ed. A.Kelly and R.Nicholson, Elsevier, 1971.
32) V. Volterra, Ann. Ecole Norm. Super., 1907, 24, 44
33) P.M. Kelly and G. Pollard, Acta Metall., 1969, 17, 1005.
34) J.S. Bowles and C.M. Wayman, Metall. Trans., 1972, 3, 1113.
35) N. Nakanishi, Prog. in Materials Science, 1980, 24, 143.

36) C.S. Barrett, Trans. J.I.M., 1976, 17, 465.
37) M. Cohen, G.B. Olson and P.C. Clapp, Proc. Int. Conf. on Martensitic
    Transformations, ICOMAT 1979, Massachusetts, p.1.
38) R.F. Bunshah and R.F. Mehl, Trans. AIME, 1953, 197, 1251.
39) K. Mukherjee, Trans. TMS-AIME, 1968, 242, 1495.
40) J.W. Christian and K.M. Knowles, "Solid-Solid Phase Transformations",
    (Eds. H.I. Aaronson et al.), 1981, TMS-AIME, Warrendale,
    Pennsylvania, p.1175.
41) G.F. Bolling and R.H. Richman, Acta Metall., 1965, 13, 745.
42) J.W. Christian, "Physical Props. of Martensite and Bainite", ISI
    spec. rep. 93, London, 1965, p.1.
43) A.B. Greninger and A.R. Troiano, Trans. AIME, 1949, 185, 590.
44) J.W. Christian, "Decomposition of Austenite by Diffusional Processes",
    (Eds. V.F. Zackay and H.I. Aaronson), 1962, Interscience,
    New York, p.371.
45) A.B. Greninger and A.R. Troiano, Trans. AIME, 1949, 185, 590.
46) E.S. Machlin and M. Cohen, Trans. AIME, 1951, 191, 1091.
47) E.S. Machlin and M. Cohen, Trans. AIME, 1952, 194, 1201.
48) J.K. Mackenzie, Aust. J. Phys., 1957, 10, 103.
49) G.B. Olson and M.Cohen, "Solid-Solid Phase Transformations",
    (Eds. H.I. Aaronson et al.), 1981, TMS-AIME, Warrendale,
    Pennsylvania, p.1209
50) B.A. Bilby, Rep. Conf. on Defects in Cryst. Solids, The Physical
    Society, London, 1955, p.263.
51) F.C. Frank, Symposium on the Plastic Defm. of Crys. Solids,
    Office of Naval Research, Pittsburgh, 1950, p.150.
52) K.M. Knowles, Philosophical Magazine A, 1982, 46, 951.
53) W.T. Read, "Dislocations in Crystals", New York, McGraw-Hill,
    1953, p.185.
54) G. Baro and H. Gleiter, Acta Metall., 1973, 21, 1405.
55) W. Bollman, Phys. Stat. Sol. (a), 1974, 21, 543.
56) M.L. Kronberg and F.H. Wilson, Trans. AIMME, 1949, 185, 501.
57) K.J. Aust and J.W. Rutter, Trans. AIMME, 1959, 215, 820.
58) D.H. Warrington and P.Bufalini, Scr. Metall., 1971, 5, 771.
59) M.A. Fortes, Revista de fisica Quimica e Engenharia, 1972, 4a, 7.
60) V. Vitek, Scr. Metall., 1970, 4, 725.
61) D.A. Smith, V. Vitek and R.C. Pond, Acta Metall., 1977, 25, 475.
62) R.C. Pond and V. Vitek, Proc. Roy. Soc. (London), 1977, A357, 471.
63) R.C. Pond, J. de Physique, 1975, C4, 315.
64) W. Bollman, Phil. Mag., 1967, 16, 363.
65) W. Bollman, "Crystal Defects and Crystalline Interfaces", Springer
    Verlag, Berlin, 1970.
66) W.T. Read and W. Shockley, Phys. Rev., 1950, 78, 275.
67) D.G. Brandon, B. Ralph, S. Ranganathan and M.S. Wald, Acta Metall.,
    1964, 12, 813.
68) T. Schober and R.W. Ballufi, Phil. Mag., 1971, 24, 165.
69) J.P. Hirth and R.W. Ballufi, Acta Metall., 1973, 20, 199.
70) D.G. Brandon, Acta Metall., 1966, 14, 479.

71) A.H. King and D.A. Smith, Acta Cryst., 1980, $\underline{A36}$, 335.

72) R.C. Pond and D.A. Smith, 4th Bolton Landing Conf. on Grain Boundaries
   in Eng. Mats., Claitons, Baton Rouge, 1975, p.309.

73) H. Grimmer, W. Bollman and D.H. Warrington, Acta Cryst., 1974,
   $\underline{A30}$, 197.

74) D.H. Warrington and H. Grimmer, Phil. Mag., 1974, $\underline{30}$, 461.

75) W. Bollman, B. Michaut and G. Sainfort, Phys. Stat. Sol. (a),
   1972, $\underline{13}$, 13.

76) A.P. Sutton and V. Vitek, Phil. Trans. R. Soc. Lond., 1983, $\underline{309}$, 1.

77) M.R. Spiegel, "Vector Analysis", Schaums outline series, McGraw Hill,
   New York, 1959.

78) F. Ayres, "Matrices", Schaums Outline series, McGraw Hill,
   New York, 1962.